PRAISE FOR THIS BOOK

"This is a well-written and comprehensive text."
—Michael D. Biderman, *University of Tennessee at Chattanooga*

"The authors provide a masterful and fluid overview of confirmatory factor analysis that will guide readers to the best practices whether conducting their own research or evaluating the research of others."
—John Hoffmann, *Brigham Young University*

"Roos and Bauldry lucidly set out foundations of confirmatory factor analysis (CFA) as applied in the assessment and construction of scales. Beginning with model specification, they discuss identification, estimation, and assessment of CFA models, before developing extensions to assessing measurement invariance and categorical (rather than quantitative) indicators."
—Peter V. Marsden, *Harvard University*

"*Confirmatory Factor Analysis* is well written and easy to read. The book covers the essentials necessary for understanding and using CFA. It is appropriate for graduate students and professors new to this analysis approach."
—Jerry J. Vaske, *Colorado State University*

Confirmatory Factor Analysis

QUANTITATIVE APPLICATIONS IN THE SOCIAL SCIENCES

SERIES: QUANTITATIVE APPLICATIONS IN THE SOCIAL SCIENCES

Series Editor: Barbara Entwisle, Sociology, The University of North Carolina at Chapel Hill

Confirmatory Factor Analysis

J. Micah Roos
Virginia Polytechnic Institute and State University

Shawn Bauldry
Purdue University

Quantitative Applications in the Social Sciences, Volume 189

Los Angeles | London | New Delhi
Singapore | Washington DC

Los Angeles | London | New Delhi
Singapore | Washington DC

FOR INFORMATION:

SAGE Publications, Inc.
2455 Teller Road
Thousand Oaks, California 91320
E-mail: order@sagepub.com

SAGE Publications Ltd.
1 Oliver's Yard
55 City Road
London EC1Y 1SP
United Kingdom

SAGE Publications India Pvt. Ltd.
B 1/I 1 Mohan Cooperative Industrial Area
Mathura Road, New Delhi 110 044
India

SAGE Publications Asia-Pacific Pte. Ltd.
18 Cross Street #10-10/11/12
China Square Central
Singapore 048423

Acquisitions Editor: Helen Salmon
Product Associate: Yumna Samie
Production Editor: Megha Negi
Copy Editor: QuADS Prepress Pvt. Ltd.
Typesetter: Integra
Indexer: Integra
Cover Designer: Candice Harman
Marketing Manager: Victoria Velasquez

Printed in the United States of America

ISBN: 978-1-5443-7513-7

This book is printed on acid-free paper.

21 22 23 24 25 10 9 8 7 6 5 4 3 2 1

CONTENTS

SERIES EDITOR'S INTRODUCTION

Measurement connects theoretical concepts to what is observable in the empirical world. As such, it is fundamental to all social and behavioral research. In this volume, Professors Micah Roos and Shawn Bauldry introduce and thoroughly explain a popular approach to measurement: confirmatory factor analysis (CFA). As the authors explain, CFA is a theoretically informed statistical framework for linking multiple observed variables to latent variables that are not directly measurable. Compared with more exploratory approaches such as principal components analysis, exploratory factor analysis, and exploratory latent class analysis, CFA tests explicit hypotheses about measurement and measurement model structures.

Confirmatory Factor Analysis targets readers well versed in linear regression and comfortable with some matrix algebra. Professors Roos and Bauldry begin with the basics, defining terms, introducing notation, and showing a wide variety of hypotheses about how latent and observed variables might be related (e.g., multiple factors, correlated errors, and cause vs. effect indicators). They proceed to a thorough treatment of model estimation, followed by a discussion of measures of fit. Readers will appreciate their practical advice—for example, what to watch out for when implementing empirical methods of model identification and how to interpret the various fit indexes. They will also value the clear explanations of multiple groups and multiple-indicator multiple-cause (MIMIC) approaches to measurement invariance. Most of the volume focuses on measures that approximate continuous variables, but Professors Roos and Bauldry also devote a chapter to categorical indicators.

In the course of discussing the ideas and procedures associated with a particular topic, each chapter develops a different example (sometimes two) covering topics as diverse as racist attitudes, theological conservatism, leadership qualities, psychological distress, self-efficacy, beliefs about democracy, and Christian nationalism drawn mainly from national surveys, including the American National Election Survey, the General Social Survey, the National Survey on Drug Use and Health, the public version of the National Longitudinal Survey of Adolescent to Adult Health, the World Values Survey, and the Values and Beliefs of the American Public Survey, but also including a series of experimental studies exploring the role of subordinate behavior in explaining the gender gap in leadership. Data to replicate the examples are available on a companion website, along with codes in R,

Stata, and Mplus, that readers may also want to use as templates for their own work.

The website can be found at **study.sagepub.com/researchmethods/qass/ roos-confirmatory-factor-analysis**.

Several features of the volume will appeal to teachers as well as learners of CFA. For example, Chapter 4 provides a simple but effective demonstration of how fit indices can go wrong. An appendix to the volume discusses the measures of reliability and the parameter estimates from CFA that readers can use to calculate more general measures of reliability than Cronbach's alpha. Readers already experienced with CFA will appreciate the careful treatment of some more advanced topics such as the model-implied instrumental variable estimator as an alternative to the maximum likelihood (ML) and robust ML estimators in Chapter 3, the discussion of multiple groups and MIMIC approaches to measurement invariance in Chapter 5, and the comparison of a weighted least squares and ML estimators when fitting measurement models with categorical indicators in Chapter 6.

Finally, this volume replaces J. Scott Long's excellent *Confirmatory Factor Analysis: A Preface to LISREL*, which was published almost four decades ago. Not only is the current volume comprehensive, including topics such as categorical indicators not covered in the earlier volume, it is less tied to a particular software platform. When I checked with him, Professor Long was pleased to pass the torch to Professors Roos and Bauldry. *Confirmatory Factor Analysis* is a worthy successor.

—Barbara Entwisle
Series Editor

PREFACE

Research across the social and behavioral sciences often involves working with concepts that can be difficult to measure. For instance, concepts such as self-efficacy, racial resentment, religiosity, class *habitus*, and nationalist sentiment have no agreed-on metrics or single items (e.g., survey questions) that capture the totality of the concepts with precision and free from measurement error. In many cases, social and behavioral scientists have multiple measures of a concept, such as self-efficacy, with each capturing different facets of the concept and containing different amounts of measurement error. These multiple observed variables can be considered indicators or measures of a latent (or unobserved) variable that represents the underlying concept of interest to the analyst.

Confirmatory factor analysis (CFA) provides a statistical framework for linking multiple observed indicators to their underlying latent variables. In general terms, CFA works by assessing the shared covariance among a set of observed variables that is typically assumed to be caused by the latent variables. CFA differs from other statistical approaches to handling latent variables in that (1) it requires the specification of a measurement model in advance as opposed to more data-driven techniques such as exploratory factor analysis (EFA), principal components analysis, and cluster analysis; and (2) it specifies latent variables as continuous rather than categorical as in latent class analysis (LCA) and latent profile analysis. CFA also shares a close affinity with item-response theory (IRT) for which the observed indicators are categorical variables (see Chapter 6 for a discussion). Among these various statistical approaches to measurement, CFA stands out as particularly well-suited to testing measurement model structures and exploring measurement properties of individual items and scales.

CFA is embedded in the larger structural equation modeling (SEM) framework that permits the specification of statistical models incorporating latent variables as well as multiple endogenous observed or latent variables (Bollen, 1989). As such we are able to rely on statistical and psychometric research on the broader class of structural equation models to inform our understanding of and recommendations for applied work in CFA. In addition, in many cases practitioners using CFA will treat it as a first step in a broader analysis that embeds a measurement model within a larger structural equation model to analyze, for instance, predictors and/or outcomes of latent variables. Where appropriate, we point out connections with structural equation models in our discussions of various aspects of CFAs, and we also refer readers interested in learning more to the various sources that provide more detail about the broader SEM framework.

This book is intended for intermediate to advanced graduate students and researchers in the social and behavioral sciences. We assume that readers will be familiar with linear regression and related statistical models. For readers who would like a refresher in this area, there are many excellent primers available (e.g., Fox, 2016; Gujarati, 2018; Lewis-Beck and Lewis-Beck, 2015). For Chapter 6, an understanding of models for categorical outcomes will be beneficial. An excellent monograph covering this material is Long's (1997) *Regression Models for Categorical and Limited Dependent Variables*. Readers interested in comparing CFA with alternative approaches to working with latent variables will find a number of other books in the Quantitative Applications in the Social Sciences (QASS) series quite valuable. Among others, these include Finch (2020) for EFA, Schuur (2011) and Ostini and Nering (2006) for different forms of IRT, and McCutcheon (1987) for a classic treatment of LCA. Readers interested in learning more about structural equation models will find a number of QASS books focused on different aspects of the framework: for nonrecursive structural models, see Paxton et al. (2011); for mediation analysis, see Iacobucci (2008); for growth modeling, see Preacher et al. (2008); and for multilevel structural equation models, see Silva et al. (2019).

Example code as well as data (variously in R using lava, Stata, and Mplus) is provided by chapter on the accompanying website at **study.sagepub .com/researchmethods/qass/roos-confirmatory-factor-analysis**.

ACKNOWLEDGMENTS

SAGE and the authors are grateful for feedback from the following reviewers in the development of this text:

Michael D. Biderman, *University of Tennessee at Chattanooga*
Kenneth A. Bollen, *University of North Carolina at Chapel Hill*
John Hoffmann, *Brigham Young University*
William G. Jacoby, *Michigan State University*
J. Scott Long, *Indiana University*
Peter V. Marsden, *Harvard University*
Vijayan K. Pillai, *University of Texas at Arlington*
Jerry J. Vaske, *Colorado State University*

ABOUT THE AUTHORS

J. Micah Roos is an associate professor of sociology at Virginia Polytechnic Institute and State University. His research interests include knowledge, science, religion, culture, stratification, measurement, racial attitudes, and quantitative methods. He ties these interests together through a quantitative, measurement approach to the sociology of knowledge and culture, with a focus on stratification along the early life course. Another strand of his work involves applying techniques in confirmatory factor analysis to the problem of differential item functioning or item-level bias. He holds a PhD in sociology from the University of North Carolina at Chapel Hill.

Shawn Bauldry is an associate professor in the Department of Sociology at Purdue University. He received a PhD in sociology and an MS in statistics from the University of North Carolina at Chapel Hill in 2012 and has previously taught at the University of Alabama at Birmingham. His research in applied statistics primarily focuses on the development of structural equation models. His research in sociology explores interrelationships between socioeconomic resources and health over the life course and across generations. His work has appeared in a variety of outlets, including *Sociological Methodology, Sociological Methods & Research, Psychological Methods, Social Forces*, and *Journal of Health and Social Behavior.*

CHAPTER 1. INTRODUCTION

1.1 Latent and Observed Variables

Social and behavioral science theories often involve abstract concepts that can be difficult to measure. Rich concepts such as social capital, racial animus, depressive symptoms, or self-efficacy are poorly approximated by single survey or questionnaire items. In many cases, analysts instead rely on multiple items to capture a concept. For instance, rather than a single question such as "During the past week, how often did you feel sad?" to capture depressive symptoms, researchers may ask several questions covering different facets of depressive symptoms, such as "During the past week, how often did you feel that everything you did was an effort?" and "During the past week, how often was your sleep restless?." These items, or observed variables, can be considered measures (or indicators) of an underlying, unobserved latent variable that represents the concept of interest.

Classical measurement theory (Lord et al., 1968) provides a framework for thinking about the relationship between observed indicators or measures and underlying latent variables. In this framework, an observed variable (x) includes two components: (1) a true component (τ) that reflects the concept of interest and (2) an error component (ε) that reflects other sources of variation in the observed variable. This relationship is represented by the simple equation $x = \tau + \varepsilon$. For instance, some component of variation in the question "During the past week, how often did you feel sad?" captures depressive symptoms and some component reflects measurement error. The key challenge, then, lies in accounting for or minimizing the presence of measurement error in the observed variable when analysts are interested in the underlying latent variable. The specification of a measurement model via confirmatory factor analysis (CFA) provides one solution to this challenge.

1.2 Reliability and Validity

Psychometricians have distinguished two measurement properties, reliability and validity, of observed indicators of latent variables (or combinations of indicators, such as scales). Drawing on the often-used darts analogy, a reliable measure is one in which repeated measurements result in quite similar responses (i.e., a cluster of darts all hitting around the same point). In contrast, a valid measure is one in which the measure captures the under-

1

lying concept that is the target of measurement. These two measurement properties are distinct in that a measure may be reliable but not valid or vice versa.

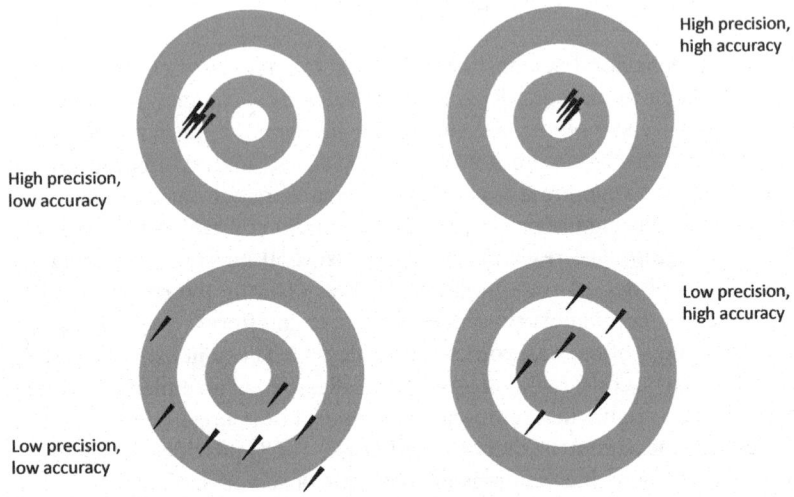

Figure 1.1: Reliability and Validity.

Figure 1.1 presents the relation between reliability and validity visually, with four bulls-eye targets struck by darts. The lower left target is low reliability, low validity—thus, the darts are not clustered well, nor is the grouping centered on the target. The upper left target is high reliability, low validity—the darts are grouped tightly but are far from the center of the target. The lower right target is the inverse, low reliability and high validity—the darts are not grouped tightly, but the grouping is centered on the target. Thus, either reliability or validity may either be high while the other is low. Last, the upper right target displays both high reliability and validity—the darts are grouped tightly and placed at the center of the target. This is the measurement ideal.

There are a number of forms of validity, including content validity, criterion validity, construct validity, and convergent and discriminant validity (Bollen, 1989). The first form, content validity, refers to whether the measures or indicators of a latent variable fully capture the domain of the latent variable. This is primarily a theoretical or conceptual concern and is not evaluated empirically. The next two forms, criterion validity and construct validity, capture the extent to which measures relate to other measures or variables that they should relate to (e.g., whether a questionnaire item measuring depressive symptoms relates to a clinical diagnosis of depression). These two forms can be empirically evaluated, but such tests rely on obtaining additional variables to do so. The fourth form, convergent and discriminant validity, concerns whether measures of a given latent variable are related to each other in the first case and whether measures of different latent variables are not related to each other in the second case. CFA provides a means of evaluating both the reliability of measures and the convergent and discriminant validity of measures.

1.3 Confirmatory Factor Analysis

CFA is a statistical model that addresses the measurement of latent variables (or factors) through the specification of a measurement model—that is, the relationships between latent variables and the observed variables (or indicators) that are thought to measure them. The fundamental idea underlying factor analysis is that a relatively small number of latent variables (or factors) are responsible for the variation and covariation among a larger set of observed indicators. The initial goal then is to identify the number of latent variables and the nature of the relationships between the latent variables and the observed indicators. Exploratory factor analysis (EFA) makes relatively few assumptions about either the number of latent variables or the relationships between the latent variables and the observed indicators and allows this information to emerge from the data. CFA, by contrast, requires theoretical or substantive knowledge to specify a measurement model in advance and then evaluates how well the specified model fits with the means, variances, and covariances of the observed indicators.

CFA and the specification of measurement models have a number of uses in the social and behavioral sciences. First, it is an invaluable tool in the psychometric evaluation of measurement instruments. For instance, an analyst may propose new questionnaire items to measure racist attitudes and could evaluate the quality of these items using CFA. Second, relatedly, CFA can be used for construct validation, particularly through assessments of convergent and discriminant validity. Third, CFA can allow analysts to explore and address various method effects, such as the effect of similar wording in the racist attitudes model. Fourth, CFA provides a framework for evaluating measurement invariance. For example, the latent variables capturing racist attitudes and the relationships between the latent variables and the indicators may vary across subgroups of the population (e.g., perhaps the measurement model differs for people living in the western United States as compared with those living in the northeast). Finally, CFA is an important first step in the specification of a broader structural equation model that allows for an analysis of structural relationships among latent variables as well as the specification of multiple endogenous variables.

Before we move to an empirical example, a note on types of indicators. Alwin (2007) details the difference between *multiple measures* and *multiple indicators*. In the first case, *multiple measures*, the intent is that identical or near-identical replicates of measures are administered more than once. In a survey, these might be the same item with word order slightly changed, or even with identical wording but administered at a later period of data collection or merely at a different point within the same period of data collection (i.e., later in the same survey interview). The degree to which item wording may vary while measures are considered identical is uncertain, and the less similar they are, the more they approach the condition of *multiple indicators* and the more complicated assessing reliability becomes. Additionally, it is generally assumed that *multiple measures* measure one and only one thing. In contrast with *multiple measures*, *multiple indicators* are not assumed to perfectly measure the same thing (a situation analysts working with secondary data often find themselves in), even though they are theoretically assumed to be measures of the same construct (an assumption we can test with CFA models). In the case of *multiple indicators*, the assumption that indicators relate only to the construct of interest is also relaxed, and it is possible that other variables may influence them (and thus, explain part of their variance). In the example that follows (and, indeed, most examples in this book), indicators are assumed to be *multiple indicators*.

To give an example, suppose an analyst is interested in characterizing the latent variables or factors that capture dimensions of contemporary racist attitudes in the United States. Social scientists have devised a set of six measures (see Table 1.1) that have been included in nationally representative surveys, such as recent waves of the American National Election Studies (ANES). The first two measures are generated from four raw survey items (two each about each group) where respondents are asked to rate on a scale of 1 to 7 for Whites and Blacks how hardworking versus lazy they are and how peaceful versus violent they are. These items are used to construct differences in the ratings for Whites and Blacks on laziness/hardworking and peaceful/violent. The second set of four measures ask respondents to indicate their agreement on a continuum from "agree" to "strongly disagree" with a series of statements about racist attitudes. Thus, an analyst working with these measures has six indicators to consider.

Background theoretical and substantive knowledge might lead an analyst to specify a measurement model for these indicators that involves two latent

Table 1.1: Item prompts and responses for racial attitudes indicators (American National Election Studies).

Next are some questions about different groups in our society. Please look, in the booklet, at a 7-point scale on which the characteristics of the people in a group can be rated.

| Lazy[a] | Where would you rate [group] in general on this scale? | [1] Hard-working to [7] lazy |
| Peaceful[a] | Where would you rate [group] in general on this scale? | [1] Peaceful to [7] violent |

The following questions are concerned with racial attitudes with responses ranging from strongly disagree (SD) to strongly agree (SA).

Try hard	It's really a matter of some people not trying hard enough; if Blacks would only try harder they could be just as well off as Whites.	[1] SD to [5] SA
Prejudice	Irish, Italians, Jews, and many other minorities overcame prejudice and worked their way up. Blacks should do the same without any special favors.	[1] SD to [5] SA
Discrimination	Generations of slavery and discrimination have created conditions that make it difficult for Blacks to work their way out of the lower class.	[1] SD to [5] SA
Deserve	Over the past few years, Blacks have gotten less than they deserve.	[1] SD to [5] SA

[a]Indicators for a measurement model were calculated as the difference in these items for Black and White groups.

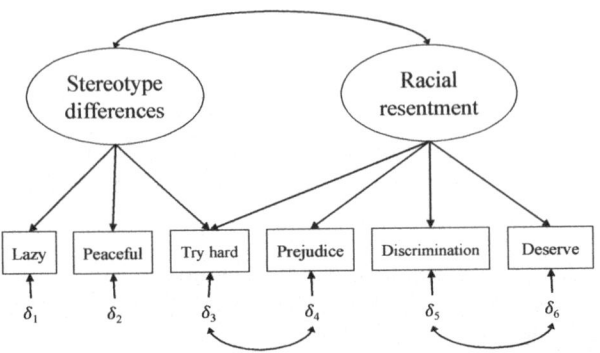

Figure 1.2: Proposed measurement model for racist attitudes.

variables. The first latent variable, which we might label "stereotype differences," is measured by the first three indicators, and the second latent variable, which we might label "racial resentment," is also measured by the third indicator as well as the remaining three indicators. Among the four indicators of racial resentment, one pair of the indicators is worded in terms of systemic oppression (e.g. "have gotten less than they deserve") and another in terms of individual actions (e.g. "if Blacks would only try harder"). These similarities in wording have the potential to introduce some amount of shared variation among the pairs of the indicators that goes beyond what is accounted for by the latent variable racial resentment. To reflect this, an analyst can specify covariances between the errors for respective indicators.

Figure 1.2 depicts a diagram that illustrates this measurement model. By convention, in diagrams for CFAs, circles or ovals indicate latent variables, squares or rectangles indicate observed variables, directed arrows indicate effects from the variable at the origin of the arrow to the variable at the destination of the arrow, and curved two-headed arrows indicate a covariance between two variables. In this particular figure, the lower curved arrows point to deltas, or δ_1 to δ_6. These Greek symbols represent measurement error or the portion of the variance in each indicator not explained by the latent variable. For this measurement model, we see the two latent variables, stereotype differences and racial resentment, represented in circles and allowed to be correlated. This specification captures the idea that these are two related but distinct dimensions of racist attitudes. We also see direct effects from the latent variables to the respective indicators that are measures of each. As noted above, the third indicator is thought to be a measure for both latent variables, and thus we see an arrow from each pointing to it. Finally, the curved two-headed arrows among the measurement errors capture the correlated errors for the two sets of indicators that measure racial resentment. This figure presents measurement models with *effect indicators*, or indicators influenced by latent variables. Other measurement forms in which indicators influence latent variables are possible (see discussion in Chapter 2).

This measurement model for racist attitudes illustrates a number of features of CFA that capture the richness of the method. We see two dimensions of racist attitudes as reflected in the two latent variables, an indicator that serves as a measure for more than one latent variable and correlated errors that account for potential method effects from similarities in the wording of the questionnaire items (this is not the only source for correlated error terms; we discuss these in greater detail in Chapter 2).

Now that we have a specified measurement model for racist attitudes, we can test the overall fit of the model and estimate the model parameters given a dataset with these measures (e.g., the ANES). As we will discuss, for many measurement models CFA allows analysts to test and assess how well the proposed model specification reproduces the means, variances, and covariances among the observed indicators. This provides a holistic assessment of the measurement model and is a key advantage of CFA as compared with EFA.

In addition to evaluating the overall fit of the measurement model, we can estimate the parameters of the measurement model. The main parameters of any CFA include "factor loadings," latent variable or factor variances and covariances if the model includes more than one latent variable, and error variances and covariances if the model specification permits as in our example. The factor loadings capture the effect of the latent variable on the indicators and are typically interpreted as regression coefficients. The factor variances capture the extent of variation across cases in the underlying latent variables. Similarly, the covariances among the factors capture the extent to which the latent variables covary with each other and can be an important component in assessing validity. For instance, if the correlation between stereotype differences and racial resentment is very close to 1.0, then we might question whether the items allow us to identify two distinct dimensions of racist attitudes. The error variances, sometimes referred to as "unique" variances, capture a combination of systematic sources of variation that affect only a given indicator and random error. We typically think of the combination of the two as "measurement error," and estimates of error variances allow us to assess the reliabilities of the indicators. Finally, the error covariances capture shared sources of variation in pairs of indicators that remain after accounting for any variation due to latent variables. We may also determine at this time whether to fix one loading per latent variable to one for identification purposes (the default in most recent software), or use other options to allow estimation of each loading (for more on these choices and model identification, see Chapter 3).

Once we have fit our measurement model to data, evaluated the overall fit of the model, and examined the parameter estimates, it is possible that we will obtain adequate model fit and reasonable parameter estimates and thus we may proceed with additional analyses or interpreting the various estimates. Alternatively, it also possible that the model fit will be inadequate or parameter estimates will be unreasonable (e.g., a negative estimate for an error variance), in which case we may explore alternative measurement model specifications.

1.4 Statistical Software and Code

Applied researchers have a number of options for statistical software to fit CFA models. Most general packages, such as R, Stata, and SAS, have the capability to fit various types of CFA models. In addition, packages designed specifically for structural equation models, such as Mplus, LISREL, and EQS, can be used for an even wider range of CFAs. Because the capabilities and syntax of the various statistical software packages continue to evolve, we have decided not to include snippets of code or output in the book. Instead we have uploaded code for multiple software packages (primarily R, Stata, and Mplus) along with data for every example in the book to a companion website at **study.sagepub.com/researchmethods/qass/roos-confirmatory-factor-analysis** to serve as a resource for readers looking to replicate the examples or use as templates for their own work. We also indicate the specific software package used for the estimates reported for the examples in the book in the notes to each table.

1.5 Outline of the Book

Chapter 2 provides a detailed introduction to model specification in CFA that includes a consideration of different types of multidimensional measurement models and the distinction between effect and causal indicators. Chapter 3 covers model identification and estimation in CFA. Model identification concerns whether it is possible to obtain unique parameter estimates and is closely connected to tests for the fit of a specified model.

Chapter 4 details model testing, both overall or global model fit and item-specific fit, and nested models. That chapter also discusses strategies for model respecification if fit is unsatisfactory. Chapter 5 examines measurement invariance—that is, whether or not items perform similarly across different groups. Assessing measurement invariance is a key part of minimizing bias in measurement. Chapter 6 introduces categorical indicators, an important extension of traditional CFA models that addresses the categorical level of measurement of many observed indicators of latent variables. Chapter 7 closes the book with concluding remarks about the broader value of CFA, a discussion of how CFA fits within the more general SEM framework, and a consideration of advanced topics with CFA along with recommendations for further study.

1.6 Further Reading

For a more in-depth examination of social concepts and their measurement, see Goertz (2020). Alwin (2007) provides an extended treatment for readers interested in learning more about reliability, validity, and measurement error with survey data. For a fuller treatment of the measurement model presented here as an empirical example, see Roos et al. (2019). For another exemplar of a measurement model with multiple latent variables, see Manglos-Weber et al. (2016).

CHAPTER 2. MODEL SPECIFICATION

The key difference between CFA and EFA lies in the opportunity to incorporate theoretical and substantive knowledge into the measurement analysis. With CFA, analysts must decide on which indicators measure which latent variables, whether measurement errors for the indicators are independent of each other or whether there are correlated errors, and, if there is more than one latent variable, which latent variables are related to each other. In addition, analysts may decide that some of the parameters in the model are fixed values or that sets of parameters are constrained to be equal. This process of defining the structure of the model and the model parameters is referred to as model specification. The specification of a measurement model is the first step in CFA.

In this chapter, we explore model specification for a variety of CFA measurement models. We begin with a simple model with a single latent variable to fix ideas and then introduce more complex models that involve multiple latent variables, indicators that are measures of multiple latent variables, method effects and correlated measurement errors, and indicators that are causes rather than effects of a latent variable.

To illustrate various measurement models, we draw on an example based on a set of measures for the concept of theological conservatism. In addition, for one of our illustrations, we also draw on two measures of religious attendance. These examples are inspired by a study using data from the 2008 General Social Survey (GSS) to examine theological conservatism among Christians in the United States (Hempel et al., 2012).

Table 2.1 provides the measures for theological conservatism as they appear in the GSS. These measures involve a mix of binary and ordinal indicators. The logic of CFA measurement model specification applies regardless of level of measurement of individual indicators. The level of measurement, however, is important to take into account when estimating model parameters. Chapter 3 provides an overview of estimators for continuous indicators, and Chapter 6 provides an overview of estimators for categorical indicators.

2.1 Forms of CFA Measurement Models

2.1.1 One Latent Variable

One of the simplest CFA measurement models is one that involves a single latent variable with a set of indicators and no correlations among the

Table 2.1: Indicator prompts and responses for theological conservatism
indicators.

Prompt	Label
Would you say you have been "born again" or have had a "born again" experience—that is, a turning point in your life when you committed yourself to Christ? (Yes/No)	reborn
Have you ever tried to encourage someone to believe in Jesus Christ or to accept Jesus Christ as his or her savior? (Yes/No)	savesoul
Which of these statements comes closest to describing your feelings about the Bible?	bible
(1) The Bible is the actual word of God and is to be taken literally, word for word. (2) The Bible is the inspired word of God but not everything in it should be taken literally, (3) The Bible is an ancient book of fables, legends, history, and moral precepts recorded by men.	
Do you believe in Hell? (Yes, definitely/Yes, probably/No, probably not/No, definitely not)	hell

measurement errors. To give a few examples, such a model could be used to assess the psychometric properties of a new set of measures (e.g., measures of student attachment to a school), could be a first step in the development of a larger measurement model with multiple latent variables (e.g., examining the relationship between student attachment to a school and parent satisfaction with a school), or could form a baseline for assessing whether the measurement model holds across subpopulations (e.g., whether measures of student attachment to a school work similarly for boys and girls).

In this example and the following examples in this chapter, we adopt the variable labeling conventions commonly used in the presentation of CFAs with observed variables denoted by x, latent variables denoted by the Greek letter ξ (xi), and measurement errors denoted by the Greek letter δ (delta). In the broader structural equation modeling (SEM) context, these labeling conventions are reserved for exogenous latent variables (i.e., variables external to the processes modeled). Endogenous latent variables, or those embedded in the processes being modeled, come with a different set of labels: y for

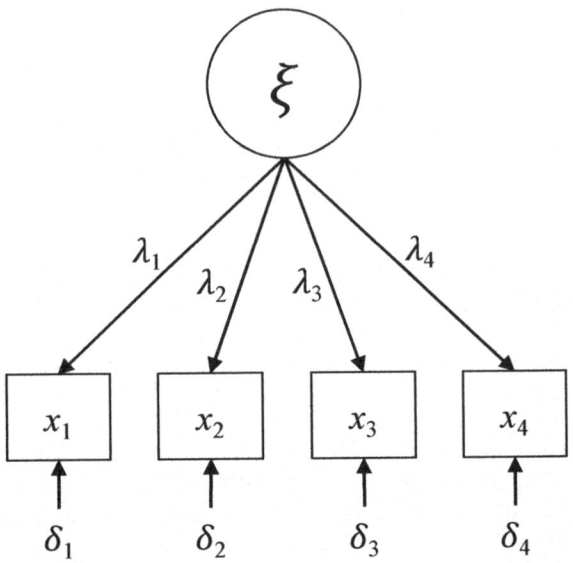

Figure 2.1: Confirmatory factor analysis measurement model with one latent variable and four indicators.

the indicators, the Greek letter η (eta) for the latent variables, and the Greek letter ε (epsilon) for the measurement errors.

Figure 2.1 depicts a measurement model with a single latent variable (ξ) and four indicators of the latent variable. Each of the four indicators (x_1 to x_4) also has a measurement error (δs), and there are no correlations among the measurement errors as evidenced by the lack of two-headed arrows connecting any of the measurement errors. Finally, the figure also includes the factor loadings, labeled with the Greek letter λ (lambda), that capture the effects of the latent variable on the indicators. This measurement model encodes the following theoretical or substantive expectations for the four indicators: (1) the four indicators are all measures of a single underlying latent variable, (2) the indicators are imperfect measures of the latent variable in that they include measurement error, and (3) the latent variable accounts for all the shared variance across the four indicators.

The measurement model illustrated in Figure 2.1 can also be represented as a system of equations as

$$x_{1i} = \alpha_1 + \lambda_1 \xi_i + \delta_{1i}$$
$$x_{2i} = \alpha_2 + \lambda_2 \xi_i + \delta_{2i}$$
$$x_{3i} = \alpha_3 + \lambda_3 \xi_i + \delta_{3i}$$
$$x_{4i} = \alpha_4 + \lambda_4 \xi_i + \delta_{4i},$$

$$(2.1)$$

where i indexes cases, αs are intercepts, λs are regression slopes or loadings, and δs are residual error terms. From this system of equations, we can see that the equation for each indicator looks much like a standard linear regression model. We have an intercept for each indicator, a regression coefficient (factor loading) giving the effect of the latent variable on the indicator, and an error term. As in a regression model, we typically assume that the mean of the errors is 0 and so the parameter of interest is the variance of the error. In contrast to a standard regression model, our predictor or independent variable is latent or unobserved. In CFA measurement models, we are often interested in the mean and variance of the latent variable(s). In total, the parameters for this model include four intercepts, four factor loadings, four measurement error variances, a mean for the latent variable, and a variance for the latent variable. As we will discuss in Chapter 3, however, we will need to fix some of these parameters to 0 or 1 in order to identify the model.

Figure 2.2 illustrates a measurement model for theological conservatism that has this form. We have the four measures of theological conservatism, *reborn*, *savesoul*, *bible*, and *hell*, represented as indicators of the latent variable for theological conservatism. In this figure, we substitute 1 in place of the first factor loading to scale the latent variable, an important component of model identification (see discussion in Chapter 3).

2.1.2 Two Latent Variables

In our first example, we examined the specification of a measurement model with only one latent variable. In many instances, however, researchers may wish to specify a measurement model with more than one latent variable. The need to include more than one latent variable can arise in several ways. First, it is possible that indicators of what are thought to be a single latent variable actually capture different dimensions of a concept and thus require more than one latent variable. For instance, one might have a set of indicators for religiosity that on reflection capture distinct domains of religiosity such as personal beliefs and engagement in religious services. In such a situation, a measurement model with latent variables for the different domains

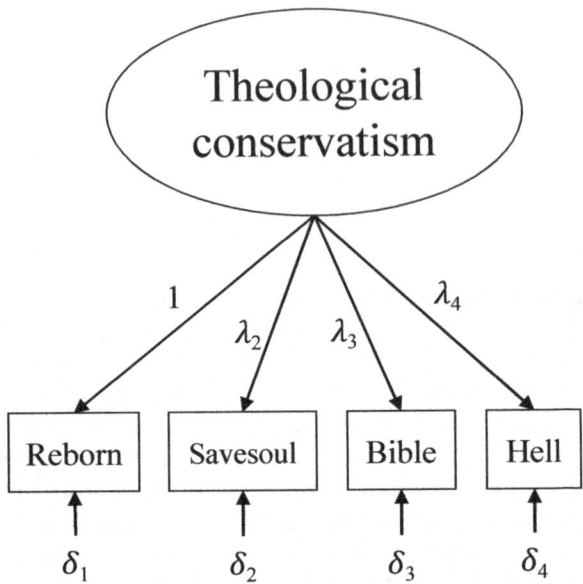

Figure 2.2: Confirmatory factor analysis measurement model for theological conservatism with four indicators.

of religiosity may perform better than a model with a single latent variable for overall religiosity. Second, the larger analytic context may involve examining the relationships between more than one latent variable. For example, a researcher may be interested in studying the relationship between student engagement and parent engagement in schools. After first developing measurement models for each separately, the researcher might combine the two into a single measurement model with both latent variables to estimate the association between student and parent engagement. Finally, in some cases latent variables may be used to capture method effects as in multitrait–multimethod (MTMM) models.

To extend the model from our first example, suppose that the analyst suspected that in fact two distinct latent variables were present that accounted for the shared variance across the four items. Figure 2.3 illustrates this model. In this model we see that x_1 and x_2 are indicators for the first latent variable and x_3 and x_4 are indicators for the second latent variable.

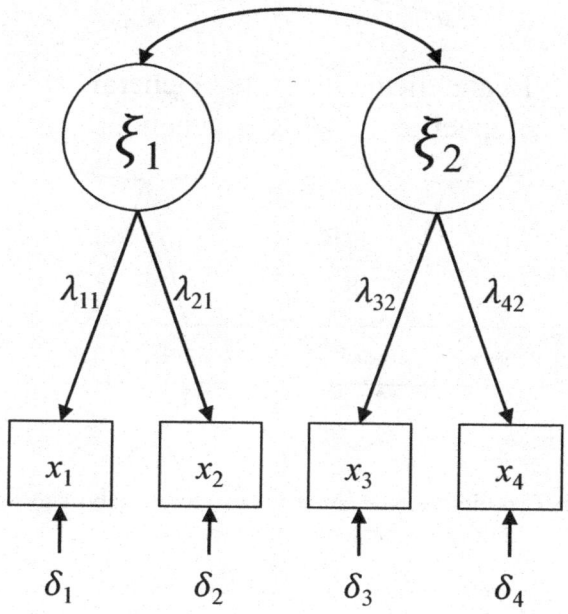

Figure 2.3: Confirmatory factor analysis measurement model with two
latent variables and four indicators.

The system of equations for this measurement model are given by

$$x_{1i} = \alpha_1 + \lambda_{11}\xi_{1i} + \delta_{1i}$$
$$x_{2i} = \alpha_2 + \lambda_{21}\xi_{1i} + \delta_{2i}$$
$$x_{3i} = \alpha_3 + \lambda_{32}\xi_{2i} + \delta_{3i} \tag{2.2}$$
$$x_{4i} = \alpha_4 + \lambda_{42}\xi_{2i} + \delta_{4i},$$

where, by convention, we provide a second subscript on the factor loadings
to indicate which latent variable has an effect on the indicator. For example,
λ_{42} represents the parameter for the factor loading for the fourth indicator
(where the 4 comes from) loading on the second latent variable (where the 2
comes from). The system of equations and the parameters are quite similar
to the previous measurement model with one latent variable. We still have
four intercepts, four factor loadings, and four measurement error variances.
For this model, we add an additional mean and variance for the second latent
variable and then also the covariance between the two latent variables.

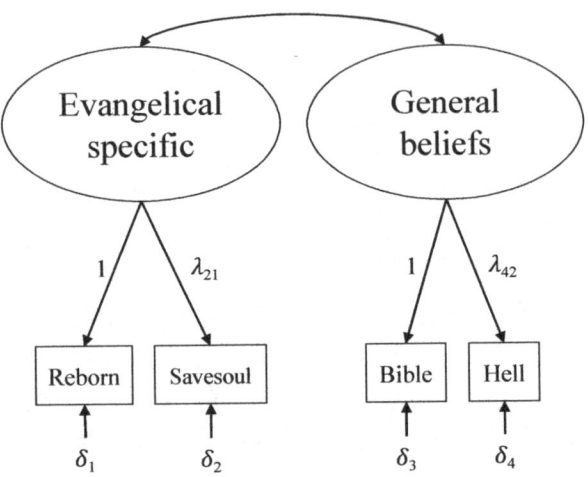

Figure 2.4: Two-dimensional confirmatory factor analysis measurement model for theological conservatism items.

Figure 2.4 illustrates an alternative measurement model for theological conservatism that instead posits that the concept consists of two dimensions: one that captures evangelical-specific beliefs and the other general beliefs regarding theological conservatism. In this case, *reborn* and *savesoul* might be best understood as indicators of evangelical-specific beliefs, while *bible* and *hell* might be best understood as indicators of more general Christian beliefs. Note that this measurement model structure implies that *reborn* and *savesoul* have a higher covariance than either has with *bible* and *hell*. To the extent this pattern among the covariances holds in the data, this measurement model will have a better fit with the data than the measurement model specifying only one latent variable. We discuss evaluating the overall measurement model fit and comparing the measurement models in Chapter 4. Some caution is warranted when working with CFAs that include only two indicators loading on a latent variable as special care must be taken to ensure such a model is identified (see Chapter 3).

2.1.3 Factor Complexity Two

It is possible that in some measurement situations an indicator may be influenced by more than one substantively meaningful latent variable. Generally, a measurement specialist creating an instrument prefers indicators that each measures only a single latent variable. In some traditions, such as Rasch

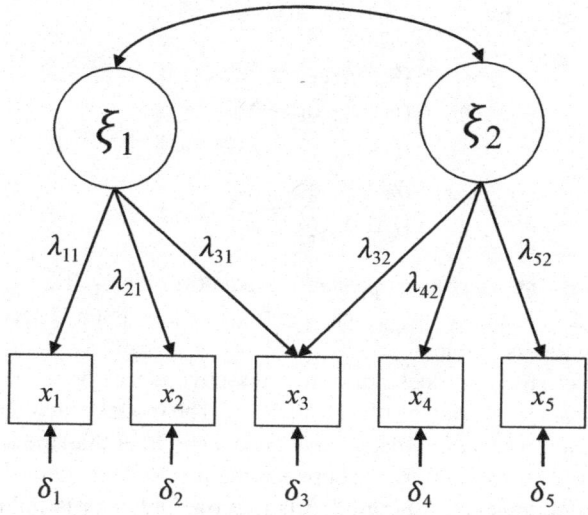

Figure 2.5: Factor complexity two confirmatory factor analysis measurement model.

modeling, it is even expected that each indicator will also have an identical loading as well as measuring only one factor. In this strict process, multidimensional items or items that measure more than one latent variable are removed (and usually considered to be poorly performing items for that reason). When performing measurement work on extant data, however, one does not always have the luxury of omitting items, and items that load on more than one latent variable can arise.

Measurement models in which one or more indicators load on two latent variables (or factors) are referred to as "factor complexity two" (FC2) models. Although rare in practice, this can be extended to multiple latent variables with factor complexity n models, where n is the greatest number of latent variables that influences a single indicator. Figure 2.5 illustrates an FC2 measurement model that involves two latent variables and five indicators. The first two indicators (x_1 and x_2) load on the first latent variable (ξ_1) and the last two indicators (x_4 and x_5) load on the second latent variable (ξ_2). The third indicator (x_3) loads on both latent variables. This set of effects for the third indicator is reflected in the system of equations for this measurement model given by

$$x_{1i} = \alpha_1 + \lambda_{11}\xi_{1i} + \delta_{1i}$$
$$x_{2i} = \alpha_2 + \lambda_{21}\xi_{1i} + \delta_{2i}$$
$$x_{3i} = \alpha_3 + \lambda_{31}\xi_{1i} + \lambda_{32}\xi_{2i} + \delta_{3i} \qquad (2.3)$$
$$x_{4i} = \alpha_4 + \lambda_{42}\xi_{2i} + \delta_{4i}$$
$$x_{5i} = \alpha_5 + \lambda_{52}\xi_{2i} + \delta_{5i},$$

where both latent variables appear in the equation for x_3. The factor loadings λ_{31} and λ_{32} capture the respective effects of the first and second latent variables on the indicator.

This model structure does not imply any covariances between the measurement errors for x_3 and the other indicators. The model structure also does not change our expectation that ξ_1 and ξ_2 covary. Other than the addition of one indicator, the sole difference between the model depicted in Figure 2.3 and the model depicted in Figure 2.5 is that one indicator is influenced by two latent variables rather than one.

Measurement models with factor complexity greater than one pose a couple of potential issues for analysts. First, as we will discuss in Chapter 3, it can be more challenging to specify models that are identified if numerous indicators load on more than one latent variable. Second, it is frequently the case in applied work that analysts make use of measurement instruments by creating a summed index of the items. Such summed indices are incapable of addressing the problem of factor complexity models, or within-item multidimensionality, as they are merely indices where each indicator is weighted equally, regardless of whether or not any one indicator is influenced by another latent variable. For this reason, when a practitioner expects their instruments to be used by analysts, they typically strive to remove the items that contribute to factor complexities greater than one from their instruments.

2.1.4   Correlated Measurement Errors

Suppose we return to our first measurement model with a single latent variable and four indicators of the latent variable. In specifying this model, we assume that all shared variance among the four indicators is due to their common dependence on a latent variable. There are a number of reasons an analyst might question this assumption. An analyst may suspect that method effects account for some of the covariance between the indicators. For instance, suppose an analyst is developing a measurement model for democracy based on four indicators and two of the indicators are ratings from the same international agency. Although some of the shared variance

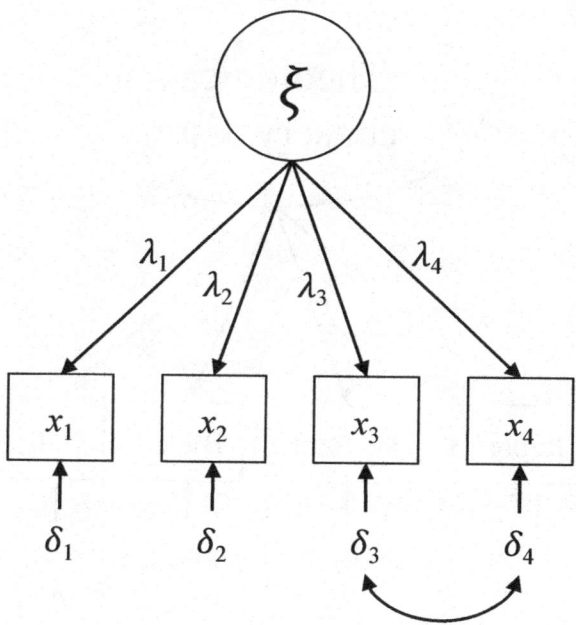

Figure 2.6: Confirmatory factor analysis measurement model with correlated measurement error.

in these two indicators reflects democracy, some shared variance may also reflect biases of the particular agency providing the ratings. Alternatively, similarly worded indicators from a survey, indicators with similar levels of reading difficulty, or indicators subject to social desirability bias, for instance, may all induce shared variance among sets of indicators and lead analysts to include correlations among measurement errors to account for the shared variance.

Figure 2.6 illustrates a measurement model that incorporates a correlation between the measurement errors for the third and fourth indicators as denoted by the two-headed arrow. The system of equations for this measurement model is the same as with Equation (2.1) and as such the model has the same set of parameters with one addition. With this model, we include the covariance between the measurement errors δ_3 and δ_4 as an additional parameter. A researcher might be tempted to allow all the measurement errors to be intercorrelated in order to capture any potential method effects or other forms of shared variance. Such a strategy, however, is not feasible

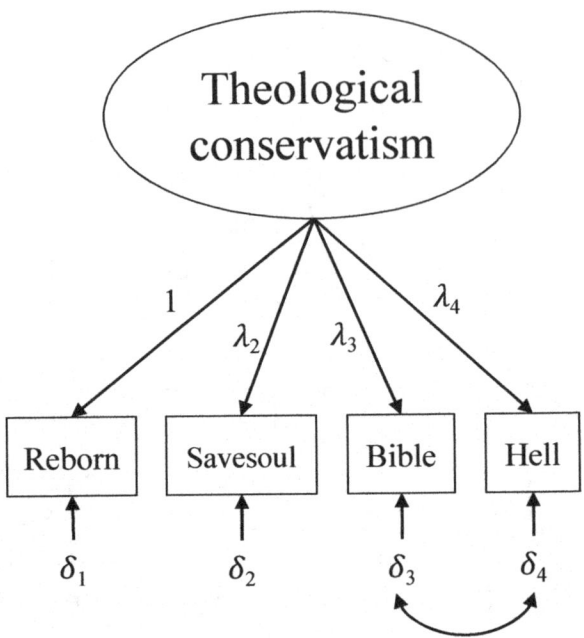

Figure 2.7: Confirmatory factor analysis measurement model for theological conservatism with correlated measurement error.

as each measurement error covariance requires a degree of freedom, and, as we will discuss in Chapter 3, researchers do not have sufficient degrees of freedom to spare for every possible measurement error covariance.

Figure 2.7 illustrates another possible measurement model for theological conservatism in which the errors for *bible* and *hell* are allowed to covary. As noted above, this model specification captures the possibility that the covariance between these indicators is not fully explained by latent theological conservatism. Such a possibility may be due to the fact that *bible* and *hell* are reverse coded or may be an artifact of the polytomous coding as compared with the dichotomous coding of *reborn* and *savesoul*. Alternately, such a possibility may reflect substantive differences in the indicators along the lines of our justification for the measurement model with two latent variable discussed above. In this case, rather than distinguishing two dimensions of theological conservatism, we might instead argue that all the indicators reflect theological conservatism but in addition *bible* and *hell* share additional covariance due to also reflecting general beliefs. This measurement model is closely related to the measurement model specifying two latent variables and, as we discuss in Chapter 4, it is not possible to distinguish empirically between the two measurement model specifications. As we discuss in Chapter 4, with a paucity of indicators (a situation many researchers will find themselves in, particularly when working with secondary data), researchers may find themselves at the limit of what the data can tell them about which model is best, and they must rely on theory in these cases.

2.1.5 Causal Indicators

The traditional CFA measurement model specifies indicators as caused by latent variables. Such indicators are commonly referred to as effect or reflective indicators. It is also possible to specify a measurement model in which indicators are causes of a latent variable rather than vice versa (Bollen and Bauldry, 2011; Bollen and Lennox, 1991). These indicators are referred to as causal or formative indicators. To give an example, socioeconomic status can be thought of as a latent variable measured by, for instance, education, income, and occupational status. If we think about these indicators, it seems unlikely that a change in socioeconomic status, the latent variable, would simultaneously lead to changes in all three indicators, as should be the case for effect indicators in which the latent variable is a cause of the indicators. Rather, a change in any of the indicators likely leads to a change in latent socioeconomic status. As such, a measurement model for socioeconomic status would be more correctly specified treating these three indicators as causal rather than effect indicators. Specifying such a model moves away from traditional CFAs and into the realm of SEM—for an extended discussion of these types of models, see Bollen and Bauldry (2011), and for a more complete treatment of the SEM framework, see Bollen (1989).

A measurement model may include a mix of causal and effect indicators as is illustrated in Figure 2.8. In this model, we have four observed measures of a single latent variable. Rather than treating all of the measures as effect indicators, as we have done in the previous models, we specify the first two measures as causal indicators and the third and fourth measures as effect indicators. The causal indicators have arrows pointing from the indicator to the latent variable to be consistent with the presumed direction of effects. In addition, the causal indicators are treated as exogenous, and thus they are allowed to be correlated and assumed to be free of measurement error (i.e., they do not have an error term associated with them). To emphasize that the causal indicators are still considered measures of the latent variable, we continue to use λ as the parameter label for the effect of the causal indicators on the latent variable. Finally, the latent variable, ξ, in this measurement

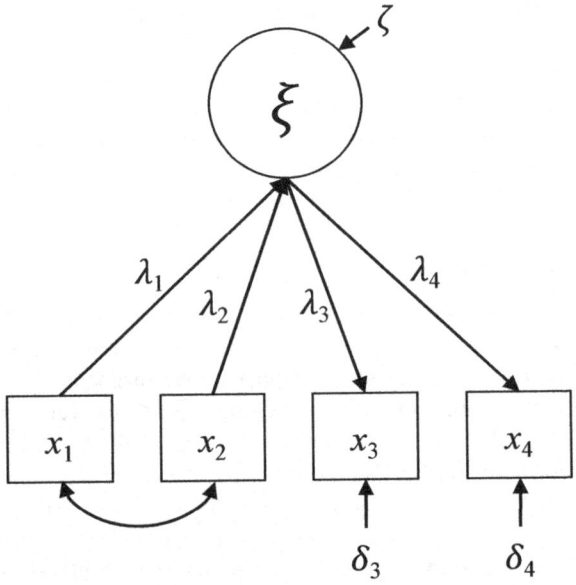

Figure 2.8: Measurement model with two causal indicators.

model is endogenous given that two of the indicators are predictors of it. This is reflected in having an error, labeled with the Greek letter ζ (zeta), pointing toward the latent variable in the figure.

The system of equations for the measurement depicted in Figure 2.8 is given as

$$
\begin{aligned}
\xi_i &= \alpha_\xi + \lambda_1 x_{1i} + \lambda_2 x_{2i} + \zeta \\
x_{3i} &= \alpha_3 + \lambda_{32}\xi_{2i} + \delta_{3i} \\
x_{4i} &= \alpha_4 + \lambda_{42}\xi_{2i} + \delta_{4i},
\end{aligned}
\tag{2.4}
$$

where the first equation in the system is for the latent variable. The parameters for this measurement model include three intercepts (one for the latent variable and two for the effect indicators), two factor loadings for the causal indicators, two factor loadings for the effect indicators, an error variance for the latent variable, two measurement error variances for the two effect indicators, and the correlation between the two causal indicators. As an endogenous variable, in this measurement model we have an intercept and an error variance for the latent variable as opposed to a mean and variance for the latent variable in a measurement model with all effect indicators. The speci-

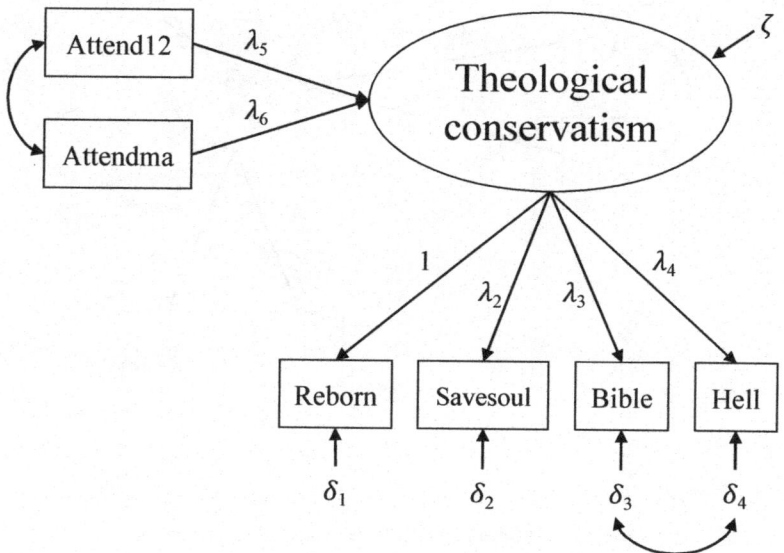

Figure 2.9: Measurement model for theological conservatism with two
causal and four effect indicators.

fication of causal indicators and the resulting shift in parameters have implications for model identification, the topic of our next chapter.

If we turn back to the example of theological conservatism, we can introduce two additional observed variables as causal indicators. For instance,
religious service attendance in adolescence (*attend12*) and mother's religious service attendance (*attendma*) are plausible indicators of theological
conservatism. Given the temporal ordering of these indicators (i.e., measured
based on adolescence as compared with adulthood for the other indicators),
it seems more reasonable to treat them as causal rather than effect indicators.
Figure 2.9 illustrates this measurement model that in addition maintains the
correlated errors for *bible* and *hell*.

2.1.6 Observed Covariates

In general, with CFA we focus on measurement models in which all the
observed variables are measures of latent variables. In some cases, however,
it can be useful to incorporate the observed variables that are not indicators
of a latent variable but instead are covariates (Bollen and Bauldry, 2011).

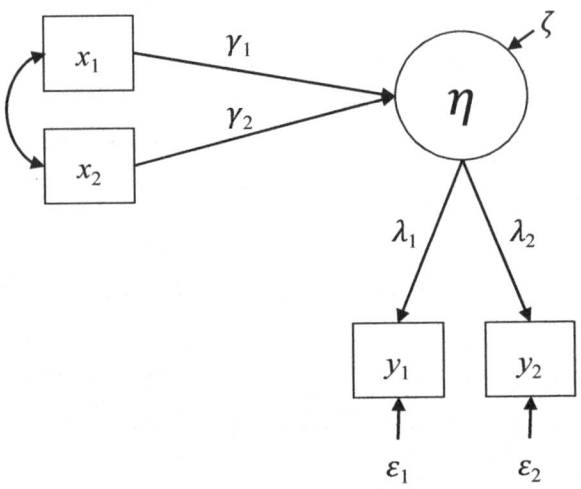

Figure 2.10: Model with two observed covariates.

This arises in the context of testing measurement invariance (see Chapter 5) and also when CFA is embedded within a larger analytic project that involves examining observed correlates, predictors, or outcomes of latent variables.

Figure 2.10 provides an example of such a model, in this case a multiple-indicator multiple-cause (MIMIC) model. In this model, we have two covariates, x_1 and x_2, that predict the latent variable. For instance, the latent variable might be religiosity as measured by two indicators and the predictors might be sociodemographic factors, such as age and gender. In this model, we label the effects of x_1 and x_2 on the latent variable using the Greek letter γ. In addition, we have switched to the endogenous latent variable notation with the Greek letter η for the latent variable, ys for the indicators of the latent variable, and εs for the measurement errors.

Attentive readers will note that despite the different notation and layout, this measurement model is identical to the model with causal indicators. The position of the covariates to the left of the latent variable rather than beneath it emphasizes that they are interpreted as predictors rather than causal indicators of the latent variable. The system of equations and the set of parameters for this model and the previous model are also identical. The identical model specifications for the previous model with causal indicators and the MIMIC model in this example highlight the critical role of theory in model specification and, ultimately, the interpretation of the estimates of CFA.

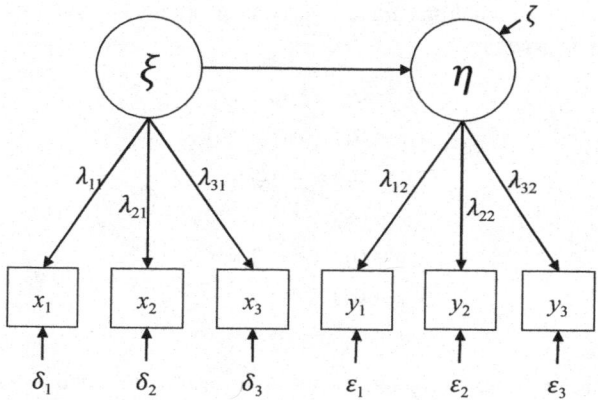

Figure 2.11: Structural equation model involving two latent variables.

2.1.7 Causal Relationship Between Latent Variables

In many cases, CFA is the first step in a larger analysis examining the relationships among a set of latent variables. If analysts are simply interested in the correlations between two or more latent variables, then the CFA framework is sufficient. If analysts, however, are interested in examining the effects of latent variables on other latent variables, then the broader SEM framework is needed. Given that such relationships are often of interest to analysts, we provide one example that embeds CFA in the more general SEM framework.

Figure 2.11 illustrates a structural equation model that involves two latent variables and a direct effect from one latent variable to the other. In this case, we have an exogenous latent variable ξ measured by three indicators x_1 to x_3 and an endogenous latent variable η also measured by three indicators y_1 to y_3. The latent exogenous variable has a direct effect on the latent endogenous variable given by γ. Previous steps in the analysis might have involved developing the measurement models for each latent variable separately and then combining them into a single measurement model with a correlation between the two latent variables.

The system of equations for the structural equation model depicted in Figure 2.11 is given by

$$\eta_i = \alpha_\eta + \gamma_1 \xi_i + \zeta$$
$$x_{1i} = \alpha_{x1} + \lambda_{11} \xi_i + \delta_{1i}$$
$$x_{2i} = \alpha_{x2} + \lambda_{21} \xi_i + \delta_{2i}$$
$$x_{3i} = \alpha_{x3} + \lambda_{31} \xi_i + \delta_{3i} \qquad (2.5)$$
$$y_{1i} = \alpha_{y1} + \lambda_{12} \eta_i + \delta_{1i}$$
$$y_{2i} = \alpha_{y2} + \lambda_{22} \eta_i + \delta_{2i}$$
$$y_{3i} = \alpha_{y3} + \lambda_{32} \eta_i + \delta_{3i},$$

where the first equation captures the relationship between the two latent variables and the remaining equations are the measurement component of the model for the two latent variables. We can see that in addition to the standard parameters from the measurement model (i.e., the indicator intercepts, factor loadings, and measurement error variances), we also have the regression coefficient for the effect of ξ on η the intercept for η and the variance of the error for η. It is worth noting that the number of parameters from this model matches the number of parameters from a model in which the two latent variables are correlated with each other and therefore there is no statistical basis for adjudicating between the structural equation model depicted in Figure 2.11 and an analogous CFA measurement model with correlated latent variables.

2.2 Conclusion

In this chapter, we introduced many of the common forms CFA measurement models can take alongside a series of examples that illustrate some of the complexity involved in specific applications. As we noted above, the first step in any CFA involves determining the concept or set of concepts to be represented as latent variables and deciding how the various observed indicators relate to the latent variables. In some contexts, this is a straightforward process as a set of indicators or measures are developed or designed to measure a specific latent variable (e.g., items to measure depressive symptoms). In other contexts, however, analysts may be working with indicators that were not initially designed to measure a specific latent variable, in which case more thought needs to given to defining a latent variable and the indicators of it. Once an analyst settles on one or more latent variables and the candidate indicators, the second step in any CFA involves specifying a measurement model along the lines we have outlined throughout this chapter.

2.3 Further Reading

Although we have discussed the most common forms of measurement models, there are a number of additional forms used in special cases. For readers interested in higher order measurement models in which latent variables are measured by other latent variables, see Brown (2015) for a discussion. For readers specifically interested in examining construct validity, an MTMM measurement model can be valuable Campbell and Fiske (1959). For an extended discussions of the distinctions between causal indicators, covariates, and composite measures, see Bollen and Bauldry (2011). For a classic exemplar of a four-indicator model, see Bollen (1982). An exemplar demonstrating model specification with correlated errors may be found in Roos (2014).

CHAPTER 3. IDENTIFICATION AND ESTIMATION

In Chapter 2, we examined a number of different types of CFA measurement models that capture a broad range of possible relationships between latent variables and indicators as well as relationships among measurement errors. In our discussion, we made reference to the issue of "identification" in connection to the need to fix certain CFA model parameters and as a limitation in the number of possible relationships that can be specified in model. In this chapter, we take up the question of model identification with a discussion of what identification refers to in the context of CFA and the standard approaches to assessing whether a given CFA model is identified.

Once a given CFA measurement model is determined to be identified, an analyst can proceed to the estimation of model parameters and an evaluation of model fit. The maximum likelihood (ML) estimator is by far the most commonly used estimator for CFA (Bollen, 1989). In addition to a discussion of identification, in this chapter we provide an overview of ML estimation and variants of ML that address departures from the multivariate normality assumption and the presence of missing data (Arbuckle, 1996; Satorra and Bentler, 1994; Yuan and Bentler, 2000). We also discuss an alternative to the ML estimator, the model-implied instrumental variable (MIIV) estimator, that addresses some of the limitations of ML estimators (Bollen, 1996).

Throughout this chapter, we draw on an example of a measurement model for leadership qualities that derives from a series of experimental studies with a total of 318 participants exploring the role of subordinate behavior in explaining the gender leadership gap (Mize, 2019). The studies include eight measures of leadership reflecting two dimensions—leadership ability and likeability (see Table 3.1). The first five items in Table 3.1 are thought to capture leadership ability, and the remaining three items are thought to capture likeability.

3.1 Identification

As we saw in Chapter 2, the specification of a measurement model involves determining the model parameters. CFA measurement models may involve three categories of parameters: fixed, constrained, and free. Fixed parameters are those that are set to specific values, most commonly 1 or 0, and thus are not estimated. In contrast, free parameters are those that may take any value and are estimated based on some data. Constrained parameters are

Table 3.1: Item prompts and responses for leadership indicators.

Please rate your group leader [name] on the following topics:										
good leader	1	2	3	4	5	6	7	8	9	bad leader
intelligent	1	2	3	4	5	6	7	8	9	unintelligent
influential	1	2	3	4	5	6	7	8	9	not influential
strong	1	2	3	4	5	6	7	8	9	weak
competent	1	2	3	4	5	6	7	8	9	incompetent
nice	1	2	3	4	5	6	7	8	9	not nice
pleasant	1	2	3	4	5	6	7	8	9	unpleasant
likeable	1	2	3	4	5	6	7	8	9	not likeable

also estimated but are subject to constraints that limit the values they may take (e.g., an equality constraint in which two factor loadings are required to be equal). In the context of CFA, identification refers to whether or not it is possible to find unique estimates for all the free and constrained parameters (i.e., unknowns) of a given model based on the information in the covariance matrix and the vector of means of the observed variables (i.e., knowns). If it is not possible, then the model is not identified. If there is exactly enough information for all of the parameters, then the model is just-identified. Finally, if there is more than enough information needed for all of the parameters, then the model is overidentified, and it is possible to test the overall fit of the model with the data (see Chapter 4). In many cases, if a model is either just-identified or overidentified, we simply refer to the model as identified.

3.1.1 Scaling Latent Variables

The first step in CFA model identification involves setting the scale for any latent variables in the model. By definition, latent variables are unobserved and thus have no defined units of measurement. There are two common approaches to setting the scale for a latent variable. The first approach is to set the metric for a latent variable to be equivalent to one of the indicators of the latent variable by fixing the factor loading for that indicator to 1 and its intercept to 0. The second approach is to fix the variance of the latent variable to 1 and the mean of the latent variable to 0. Both approaches are equivalent in terms of overall model fit but have different benefits in different research contexts. The first approach permits researchers to obtain unstandardized estimates that are important for assessing measurement invariance (see Chapter 5) and scale reliability. The second approach

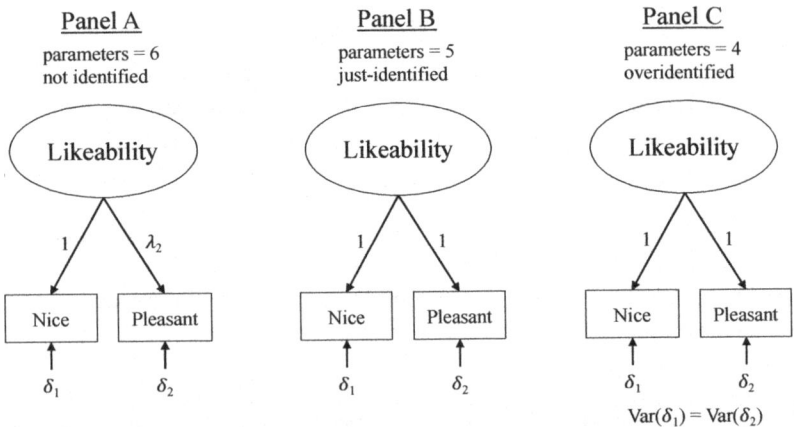

Figure 3.1: Three measurement models involving two indicators.

generates standardized estimates that can be useful when the indicators themselves do not have particularly meaningful metrics.[1]

3.1.2 Establishing Model Identification

Setting the scale for latent variables by fixing parameters is a necessary but not sufficient condition for CFA model identification. We still need to determine whether there is enough information to obtain unique estimates for all the free and constrained parameters. To gain intuition about this, let us consider the three indicators of likeability, nice, pleasant, and likeable, discussed above. First, suppose we just have two of the three indicators. With two indicators, we have the mean of each indicator, the variance of each indicator, and a covariance between the two indicators for five knowns.

Figure 3.1 illustrates three possible models for the two indicators of likeability. The first model depicted in Panel A treats *nice* as the scaling or referent indicator with its factor loading fixed to 1 and its intercept fixed to 0. This leaves the factor loading and intercept for the second indicator, error variances for both indicators, and the mean and variance of latent likeability as free parameters. With only five knowns from the two observed variables, we see that we do not have enough information to obtain unique estimates

---

[1]Some statistical software packages default to a mix of the two approaches with, for instance, the factor loading of the first indicator fixed to 1 and the mean of the latent variable fixed to 0.

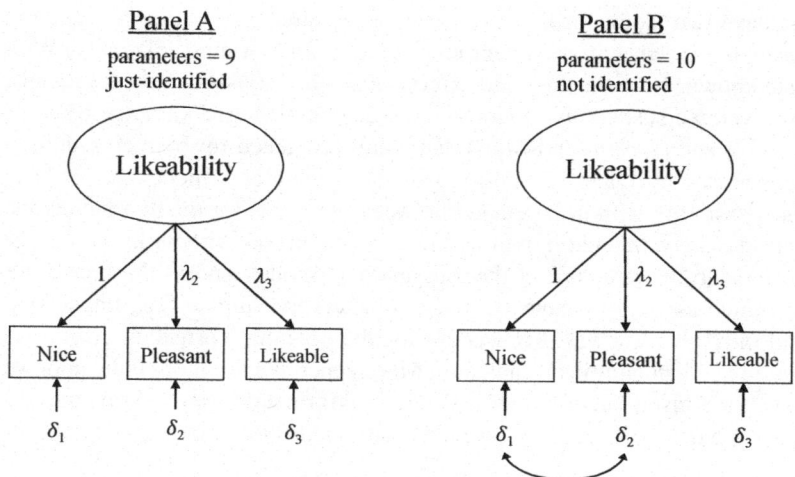

Figure 3.2: Two measurement models involving three indicators.

for each of the six free parameters. This model is, therefore, not identified, and we are unable to find estimates for the parameters. Moving to Panel B, perhaps we could make the case that *nice* and *pleasant* are equivalent indicators and have the same relationship with likeability. As such, we could fix the factor loading for pleasant to 1 in our measurement model. Now we have a model with five parameters, and with five knowns the model is just-identified. Finally, suppose we go a step further and impose an equality constraint that the error variances for the two indicators are equal to each other. In this case, our model is down to four parameters and is overidentified such that we can obtain estimates for the four parameters and also test the fit of the model against the data. As we can see, with a single latent variable and only two indicators, we need to make strong assumptions about the relationships between the indicators and the latent variable in order to specify an identified measurement model.

Now suppose we have all three indicators of likeability. With three indicators, we have the mean of each indicator, the variance of each indicator, and three covariances for nine knowns. Figure 3.2 illustrates two measurement models for the three indicators of likeability. In Panel A, we see a model in which the first indicator, *nice*, is used to scale the latent variable with its factor loading fixed to 1 and its intercept fixed to 0. As opposed to the model with two indicators, we see that the other factor loadings are free to be estimated, and there are no constraints on any of the parameters. For this model, we have two factor loadings, two intercepts, three error variances, and the mean and variance of the latent variable for a total of nine unknowns. With nine knowns from the three indicators, this model is just-identified. Perhaps, however, we suspect that there is some degree of shared variance between the *nice* and *pleasant* indicators that is not accounted for by their common dependence on latent likeability. Panel B illustrates a model that reflects this possibility with a correlation between the errors for the two indicators. Unfortunately, the addition of another free parameter without making other changes to the structure of the measurement model renders the model not identified as we now have 10 free parameters to estimate (i.e., unknowns) and only nine knowns. To account for the potential correlated errors, we would need an additional indicator. More generally, the more indicators we have for a given latent variable, the more information we can draw on, and therefore the more flexibility we have with model specification.

3.1.3   General Rules for Identification

With more complicated models, a simple comparison of the number of observed means, variances, and covariances (i.e., knowns) and the number of free and constrained parameters to be estimated (i.e., unknowns) is insufficient for establishing whether the model is identified. More precisely, having at least as many knowns as unknowns is a necessary, but not sufficient, condition for establishing measurement model identification. Nonetheless, it is a good first step. The general rule is formally known as the *t*-rule and can be stated as

$$t \leq \frac{1}{2}(q)(q+3), \tag{3.1}$$

where *t* is the number of free and constrained parameters to be estimated and *q* is the number of observed indicators.

A number of general rules have been established that provide sufficient conditions for model identification (Bollen, 1989). Among the most useful for commonly encountered models are the three-indicator rule and the two-indicator rule. The three-indicator rule stipulates that a measurement model in which each latent variable has at least three indicators with nonzero factor loadings and no correlated errors is identified. For models involving more than one latent variable, the two-indicator rule stipulates that the measurement model is identified if each latent variable has at least two indicators with nonzero factor loadings and no correlated errors. These two rules for CFA measurement model identification are sufficient, but not necessary, conditions for identification. In other words, measurement models that do not meet these conditions may still be identified depending on the presence of additional constraints or more observed indicators.

3.1.4 Empirical Tests for Identification

When examining more complex model specifications involving parameter constraints and correlated errors, it can be difficult to work out whether the rules apply to a given CFA measurement model, and it is possible to specify measurement models that are not covered by any of the general rules. Fortunately, an alternative, albeit an imperfect one, is available with empirical tests for model identification.

Empirical tests for identification emerge as a byproduct of maximum likelihood estimation and thus require no additional effort from the analyst. There are a few versions of these types of tests for identification, but the most common involves checking whether the information matrix, a matrix needed for calculating standard errors, is singular (or noninvertible). A singular information matrix is indicative of a model that is not identified (Rothenberg, 1971). Standard software for fitting CFAs provide warnings whenever a singular information matrix arises. Analysts can be reasonably confident in the absence of such a warning that the given specified CFA is identified.

As mentioned above, however, empirical tests of identification are imperfect in a couple of ways. First, these tests of identification are for *local* rather than *global* identification. This means that the tests only ensure that unique parameter estimates exist in a neighborhood of the parameter space rather than across the entire parameter space. A well-known example of this occurs with a measurement model with a single latent variable, with three indicators for the latent variable, and with the latent variable scaled by setting its variance to 1. If we used this model specification for likeability and fit it to the experimental data, we could obtain two different sets of estimates for the factor loadings. In this case, one set has all positive values around 1.5 and the other set is the exact inverse with all negative values around 1.5. Which set of factor loadings we obtain depends on the starting values (discussed below). The parameter estimates for this model are unique only in the neighborhood of positive values (or negative values) of the parameter space rather than across the entire parameter space. In a technical sense, this model is locally identified and passes the test of the singularity of the information matrix, but it is not globally identified. Since the nonunique estimates of the factor loadings are exact inverses of each other, the lack of global identification is not considered a problem since it does not affect interpretation—for example, an analyst can reverse the direction of the label of the latent variable to unlikeability if the negative set of factor loadings emerges. This is not always the case, though, and analysts run a slight risk of encountering a less benign version of a locally identified but not globally identified CFA model when relying on empirical tests of identification.

The first limitation of empirical tests of identification can lead analysts astray in accepting a set of estimates as unique when they are not. The remaining limitations lead to the opposite problem, assuming a model is not identified when in fact it may be. A singular information matrix is an indication that a model is not identified, but there are other reasons an information matrix may be singular that are not related to identification. In other words, a warning of a singular information matrix is an indication of a problem that may or may not involve model identification. In addition, there is the issue of *empirical underidentification* in which a given measurement model is identified but specific parameter estimates from a given source of data lead to a not identified model. For instance, suppose a mistake was made in one of the leadership experiments and unbeknownst to the researcher the indicator *likeable* for latent likeability was replaced by random noise due to a data entry error. In such a case, it is possible that the estimate for that factor loading would be essentially 0 and the model would be flagged as not identified because empirically with these particular data it would look like a two-indicator rather than a three-indicator measurement model.

3.1.5   Algebraic Methods for Identification

In the rare instances when empirical tests for identification fail and when the specified model is not covered by any of the established rules of identification, we need to turn to algebraic methods for determining the identification status of a model. An algebraic approach to identification involves solving for all the free and constrained parameters in a model in terms of the means, variances, and covariances of the observed variables. This process generally begins by writing the equations linking the covariance matrix and mean vector of the observed variables to the covariance matrix and mean vector implied by the specification of the model (see discussion below) and then solving the system of equations. For complex CFA models, this is often a difficult task and can be aided by computer algebra software (Bollen and Bauldry, 2010).

To give an example, suppose we want to determine the identification status of the measurement model for likeability with the latent variable scaled by fixing the variance of likeability to 1 and the mean of likeability to 0 using an algebraic approach. To simplify our example, we will focus on the observed variances and covariances for the three factor loadings and error variances and ignore the vector of means for the three intercepts. The general matrix form for the implied covariance matrix for a CFA is given by

$$\Sigma(\theta) = \Lambda \Phi \Lambda' + \Theta_\delta, \tag{3.2}$$

where $\Sigma(\theta)$ is the implied covariance matrix expressed as a function of the parameters θ, Λ is a matrix of factor loadings, Φ is the covariance matrix for the latent variables, Θ_δ is the covariance matrix for the indicator errors, and $'$ indicates a matrix transpose (see Bollen, 1989, p. 236, for a derivation of this equation). For the proposed measurement model for likeability, Equation (3.2) simplifies to

$$\Sigma(\theta) = \begin{bmatrix} \lambda_1 \\ \lambda_2 \\ \lambda_3 \end{bmatrix} [1] \begin{bmatrix} \lambda_1 \\ \lambda_2 \\ \lambda_3 \end{bmatrix}' + \begin{bmatrix} \theta_1 & 0 & 0 \\ 0 & \theta_2 & 0 \\ 0 & 0 & \theta_3 \end{bmatrix} \qquad (3.3)$$

$$= \begin{bmatrix} \lambda_1^2 + \theta_1 & \lambda_1\lambda_2 & \lambda_1\lambda_3 \\ \lambda_1\lambda_2 & \lambda_2^2 + \theta_2 & \lambda_2\lambda_3 \\ \lambda_1\lambda_3 & \lambda_2\lambda_3 & \lambda_3^2 + \theta_3 \end{bmatrix}. \qquad (3.4)$$

Now consider the variance–covariance matrix for the observed indicators. We use the Greek letter sigma (σ) to denote these variances and covariances—for example, $\sigma_1^2 = \text{var}(nice)$ and $\sigma_{12} = \text{cov}(nice, pleasant)$. If we equate Equation (3.4) to the variance–covariance matrix for the three indicators of likeability, we have

$$\begin{bmatrix} \sigma_1^2 & \sigma_{12} & \sigma_{13} \\ \sigma_{21} & \sigma_2^2 & \sigma_{23} \\ \sigma_{31} & \sigma_{32} & \sigma_3^2 \end{bmatrix} = \begin{bmatrix} \lambda_1^2 + \theta_1 & \lambda_1\lambda_2 & \lambda_1\lambda_3 \\ \lambda_1\lambda_2 & \lambda_2^2 + \theta_2 & \lambda_2\lambda_3 \\ \lambda_1\lambda_3 & \lambda_2\lambda_3 & \lambda_3^2 + \theta_3 \end{bmatrix}. \qquad (3.5)$$

Note that σ_{21} is the same as σ_{12} and thus redundant. We can form a system of six equations with six unknowns from the nonredundant elements of the two matrices and attempt to solve the system. If we are able to show that it is possible to express every parameter as a function of the observed variances and covariances, then we will have established that the specified model is identified. In this case, we find, for example, that

$$\lambda_1 = \pm \frac{\sigma_{22}}{\sqrt{\frac{\sigma_{22}\sigma_{32}}{\sigma_{21}}}}. \qquad (3.6)$$

and

$$\theta_1 = \frac{\sigma_{11}\sigma_{32} - \sigma_{21}\sigma_{22}}{\sigma_{32}}. \qquad (3.7)$$

We can derive similar expressions for all the parameters. Notice the presence of the plus/minus sign in the equation for λ_1. This indicates the two solutions for the factor loadings for this model noted above.

3.1.6   Summary of Model Identification

To summarize, in most cases, either established rules or empirical tests of identification will alert analysts to specified models that are not identified. If an analyst encounters such a model, then the analyst should consider whether adjustments to the model specification (e.g., constraining some parameters to be equal, fixing one or more parameters to specific values, or other alterations to the structure of the measurement model) are defensible based on theory or substantive knowledge. Finally, if there are concerns about the empirical tests of identification (i.e., either a failed test for a model that is believed to be identified or vice versa), then there are two options. Either the analyst can check whether the model is covered by one of the established rules of identification or can attempt to find algebraic solutions for all the parameters, likely with the aid of a computer algebra system.

3.2   Estimation

Now that we have an understanding of CFA model identification, in this section, we turn to the estimation of measurement model parameters. In the previous section, we noted that the specification of a CFA measurement model implies that certain relationships will exist among the means, variances, and covariances of the observed variables. This information is contained in what are typically referred to as the implied moment matrices or the implied covariance matrix and implied mean vector. Equation 3.2 provides a general expression for the implied covariance matrix and a similar type of equation can also be derived for the implied mean vector (Bollen, 1989).

In the context of CFA, estimation typically involves finding values for the model parameters that generate implied moment matrices that are as close as possible to the observed moment matrices (i.e., the vector of means and covariance matrix of the observed variables). For instance, in our example above with the measurement model for likeability, we found that the specified model with the variance of likeability fixed to 1 implied that the covariance between *nice* and *pleasant* is given by

$$\sigma_{21} = \lambda_1 \lambda_2. \tag{3.8}$$

Estimates for λ_1 and λ_2 when multiplied together should be as close as possible to the observed covariance between *nice* and *pleasant*. More generally, most estimators for CFA involve specifying a fitting function that minimizes the difference between the implied moment matrices and the observed moment matrices.

3.2.1 Maximum Likelihood

By far, the most commonly used fitting function for CFAs is the ML fitting function. ML is based on the principle of finding a set of parameter estimates that maximize the probability of the observing the values in the data (see Eliason, 1993, for an excellent general introduction). To do this, in the CFA context, we assume that the observed data come from a multivariate normal distribution and this allows us to derive the ML fitting function (see Bollen, 1989, pp. 131–135, for a derivation).

The ML fitting function for CFAs does not have a closed form solution, which means that the parameter estimates need to be obtained in an iterative fashion. We begin with an initial set of values for the parameters (i.e., starting values) and then use an algorithm to adjust these estimates to improve the value of the ML fitting function. Improvements in the fitting function represent parameter estimates that decrease the difference between the implied moment matrices and the observed moment matrices. If we reach a point where the adjustments to the estimates no longer lead to appreciable improvements in the fitting function, then the model is said to have *converged*, and the resulting estimates are the ML parameter estimates.

Alternatively, if at the end of a set number of iterations of the algorithm, the fitting function continues to improve, then the model has *not converged*, and any resulting parameter estimates are deemed untrustworthy. A lack of convergence can stem from a number of sources, including but not limited to (1) a misspecified model, (2) an empirically underidentified model, and (3) numerical issues due to sparse regions of the data. As a pair of simple steps to address the problem of a model that fails to converge, researchers can increase the number of iterations and input different starting values. If these steps do not work, then the researcher can inspect descriptive statistics of the observed variables to identify highly skewed variables or other unusual distributions that may contribute to sparse regions of the data. In addition, the researcher can examine estimates from the model that failed to converge to diagnose potential issues (e.g., a poorly performing indicator) and develop an alternative model (see Chapter 4).

The popularity of the ML fitting function stems from a number of desirable properties. First, it produces standard errors for the parameter estimates as a by-product of the estimation procedure. Second, the fitting function multiplied by $N - 1$ has a chi-square distribution and can be used in tests of overall model fit (see Chapter 4). Third, ML estimators, in general, are asymptotically unbiased, consistent, and asymptotically efficient, which means that in sufficiently large samples they will perform at least as well as any other estimator (provided the multivariate normality assumption holds).

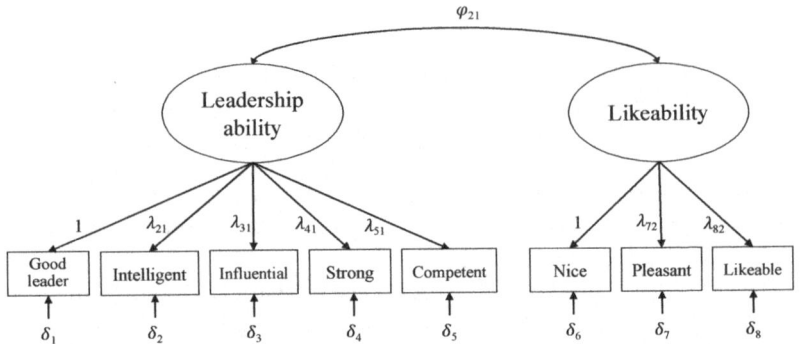

Figure 3.3: Candidate measurement model for leadership.

As an illustration, let us consider a full measurement model for leadership that specifies two latent variables and a total of eight indicators. Figure 3.3 illustrates what we might think of as a baseline measurement model for leadership with two latent dimensions (leadership ability and likeability), five indicators for the first dimension, and three indicators for the second dimension. This model specification includes 25 parameters: six factor loadings, six indicator intercepts, eight indicator error variances, two latent variable means, two latent variable variances, and one covariance between the two latent variables. In a standard statistical software package, the ML fitting function converged in 13 iterations for this model.

Table 3.2 reports the ML estimates for our baseline leadership measurement model. Setting aside for a moment the question of how well this model fits with the data overall (a topic we cover in the next chapter), we can interpret the various types of parameter estimates. First, we recall that the good leader and the nice indicators are the scaling indicators for latent leadership ability and latent likeability, respectively, and thus have fixed values of 1 for their factor loadings and 0 for their intercepts. The remaining indicators have estimated factor loadings that all range around 1. These are unstandardized estimates and are interpreted as a 1-unit increase in the latent variable is associated with roughly a 1-unit increase in each of the indicators.[2] Or, to be more precise, a 1-unit increase in leadership ability is associated with a 1.23-unit increase in the intelligent indicator, a 1.27-unit increase in the influential

---

[2]In some settings, it is common to examine and report standardized factor loadings, which are readily available in all statistical packages that fit CFA models.

Table 3.2: Maximum likelihood estimates for leadership measurement model.

	Panel A			
	Loadings	Intercepts	Error Var	R^2
Leadership ability				
Good leader	1 (—)	0 (—)	1.28 (0.11)	0.48
Intelligent	1.23 (0.09)	−0.89 (0.31)	0.88 (0.09)	0.67
Influential	1.27 (0.11)	−0.15 (0.38)	2.00 (0.18)	0.49
Strong	1.17 (0.09)	−0.21 (0.29)	0.62 (0.07)	0.73
Competent	1.19 (0.09)	−0.70 (0.30)	0.80 (0.08)	0.68
Likeability				
Nice	1 (—)	0 (—)	0.24 (0.03)	0.90
Pleasant	1.01 (0.03)	−0.14 (0.10)	0.12 (0.03)	0.81
Likeable	1.00 (0.03)	−0.03 (0.13)	0.50 (0.05)	0.95

	Panel B	
	Means	Variances
Leadership ability (LA)	3.22 (0.09)	1.20 (0.18)
Likeability (L)	3.69 (0.09)	2.19 (0.19)
Cov(LA, L)	0.93 (0.12)	

Note: Unstandardized estimates with standard errors are in parentheses. Estimates were obtained from Stata. Var = variance.

indicator, and so on. The standard errors for the estimates appear in parentheses. Invoking the multivariate normality assumption, we can assess the statistical significance of the factor loadings. We see that ratio of the estimates of the factor loadings to their standard errors are all greater than 1.96 and thus statistically significant at a conventional level of $\alpha < 0.05$. In other words, the two latent variables have statistically significant associations with all of their respective indicators. If a latent variable has a minimal association with a specified indicator or has an association in the opposite direction than expected, then that provides evidence that the specified indicator is not a good measure of the latent variable or that the measurement model more broadly is misspecified.

Next we turn to the estimates of the intercepts and error variances. For the intercepts, we see a range of estimates that can be interpreted as the value the indicator takes when the latent variable takes a value of 0 (i.e., the same interpretation as in a bivariate regression model). The indicator intercepts, however, are not often interpreted in most CFA contexts. We also see a range of values for the error variances. As with the intercepts, the error variances are not often directly interpreted. Instead, they are used in the calculation of the amount of variance in the indicators explained by the latent variables, referred to as the *communalities* or R^2s of the indicators. The final column of Panel A in Table 3.2 reports the R^2s (communalities) for each indicator. We see R^2s that range from 0.48 to 0.73 among the indicators of the latent leadership ability and 0.81 to 0.95 among the indicators of latent likeability. In the context of CFA, the R^2s for the indicators can be interpreted as a measure of reliability of the indicator. Such an interpretation can be useful, for instance, when examining the psychometric properties of a set of items for a new scale. Items with low reliabilities may be considered poor candidates for inclusion in the scale.

Finally, we can examine the means, variances, and the covariance among our two latent variables. We see that the estimated means for latent leadership ability and likeability are, respectively, 3.22 and 3.69. Since the latent variables are scaled respectively to *good leader* and *nice*, we can see that these means are on the positive side for leadership quality (see Table 3.1). Similarly, the estimates for the variances of the two latent variables are 1.20 and 2.19, respectively, and the estimate for the covariance is 0.93. It is often useful to standardize the covariance to obtain a correlation. In this case, the correlation between latent leadership ability and likeability is 0.57, which is indicative of a moderate positive association between the two dimensions of leadership.

In some cases, analysts will encounter a negative estimate for a variance (either for a latent variable or for an indicator measurement error) or an estimate of a covariance that when standardized is greater than 1 or less than -1. Such inadmissible estimates are referred to as Heywood cases and can arise from a variety of sources, such as the presence of outliers among the indicators, issues with model identification (particularly empirical underidentification), and model misspecification (Kolenikov and Bollen, 2012). If such an estimate is encountered, readers may consult Kolenikov and Bollen (2012) for diagnostics and recommendations for how to proceed.

3.2.2 Robust Maximum Likelihood

As noted above, the standard ML estimator rests on the assumption that the observed indicators follow a multivariate normal distribution. Figure 3.4 illustrates the distributions of the indicators. It is apparent that none of the indicators follow a univariate normal distribution much less a multivariate

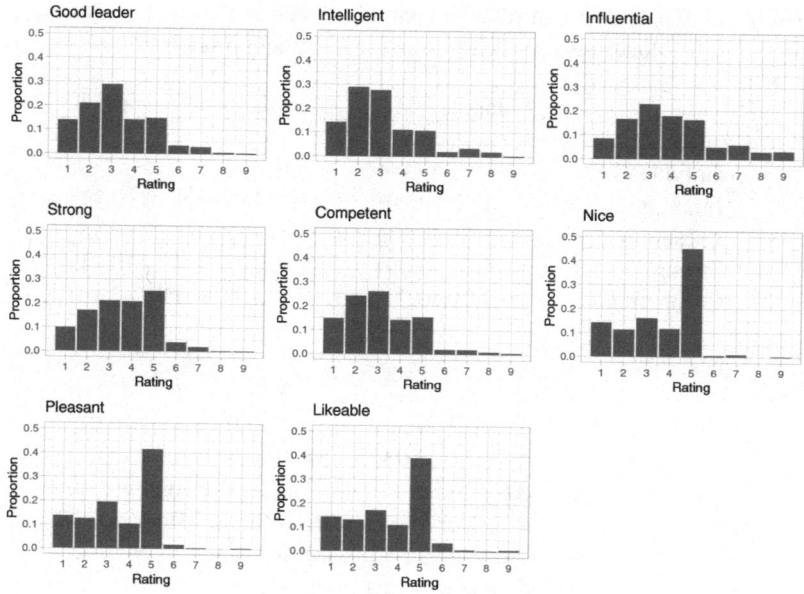

Figure 3.4: Distribution of indicators of leadership ability and likeability.

normal distribution across the indicators. Although numerous simulation studies find that the ML estimator is reasonably robust to moderate violations of the multivariate normal assumption, statisticians have worked out adjustments to the standard ML estimator to help account for observed indicators that follow nonnormal distributions. In this section, we discuss two robust ML estimators that are appropriate for interval-level or roughly continuous indicators that do not follow normal distributions. These estimators are also sometimes used for models involving categorical indicators and can perform reasonably well; however, we will examine estimators specifically designed for categorical indicators in Chapter 6.

Robust ML estimators produce ML parameter estimates with adjustments to the standard errors of the parameter estimates and the overall chi-square test statistic (discussed in the next chapter) that account for the nonnormal distribution of the data. The first robust ML estimator, labeled MLM for mean-adjusted maximum likelihood, adjusts just for the nonnormal distributions of the observed indicators. This estimator produces a corrected chi-square statistic that is often referred to as the Satorra–Bentler scaled χ^2 (Satorra and Bentler, 1994). The second robust ML estimator, labeled

Table 3.3: Comparison of robust maximum likelihood estimates for factor loadings from leadership measurement model.

	Standard ML	MLM	MLR
Leadership Ability			
Good leader	1 (—)	1 (—)	1 (—)
Intelligent	1.23 (0.09)	1.23 (0.11)	1.23 (0.11)
Influential	1.27 (0.11)	1.27 (0.12)	1.27 (0.13)
Strong	1.17 (0.09)	1.17 (0.10)	1.17 (0.11)
Competent	1.19 (0.09)	1.19 (0.10)	1.19 (0.10)
Likeability			
Nice	1 (—)	1 (—)	1 (—)
Pleasant	1.01 (0.03)	1.01 (0.02)	1.01 (0.02)
Likeable	1.00 (0.03)	1.00 (0.03)	1.00 (0.03)

Note: Unstandardized estimates with standard errors are in parentheses. Estimates were obtained from Stata. ML = maximum likelihood; MLM = mean-adjusted maximum likelihood; MLR = maximum likelihood robust.

MLR for maximum likelihood robust, adjusts for nonnormal distributions of the observed indicators and other potential violations of basic statistical assumptions (e.g., the assumption that the observations are independent and identically distributed—the iid assumption). In addition, the MLR estimator is appropriate when using ML to address missing data (discussed below).

Table 3.3 reports the estimates of the unstandardized factor loadings and standard errors from a standard ML estimator and the two robust ML estimators. We note first that the estimates of the factor loadings themselves do not change when using a robust ML estimator as opposed to a standard ML estimator. Applying the Satorra–Bentler correction results in standard errors that are slightly higher for the indicators of latent leadership ability and slightly lower for the indicators of latent likeability. To give an example, the standard error for the *intelligent* indicator is 0.09 with the standard ML estimator and 0.11 with the MLM estimator. Using the more general robust ML estimator, MLR, results in standard errors that are slightly higher than either the ML or MLM estimators for the indicators of latent leadership ability and about the same as the MLM estimator standard errors for the indicators of latent likeability.

Robust ML estimators are frequently used in practice when researchers encounter indicators with nonnormal distributions. These estimators maintain the standard ML estimates while adjusting the standard errors (and overall model fit statistic as discussed in the next chapter). As we observed in our example, the adjustments may lead to higher or lower standard errors than with the standard ML estimator.

3.2.3 Maximum Likelihood With Missing Data

In addition to nonnormally distributed data, researchers often encounter incomplete or missing data when fitting CFA measurement models. There are several contemporary approaches to addressing missing data with CFAs, including listwise deletion, the expectation-maximization (EM) algorithm, and multiple imputation. The most commonly recommended approach, however, involves a straightforward extension of the standard ML estimator that simply uses the available data for each case and in essence averages over the missing data (Arbuckle, 1996). This estimator goes by a number of different names including a full-information maximum likelihood (FIML) estimator, a direct maximum likelihood (DML), and a case-wise maximum likelihood (CML) estimator, among others. Before providing an example of this estimator, we consider first the sources and types of missing data.

Data for the observed indicators or a latent variable may be missing for a number of reasons. In some cases, data may be missing for some participants or respondents due to the research design. For instance, participants may be randomly selected to complete different batteries of items in order to save time or money. In other cases, missing data are not planned and instead due to chance (e.g., participants may unintentionally skip providing a response to some questions) or due to systematic patterns unrelated to the research design (e.g., certain types of participants might be less likely to respond to some items).

Statisticians distinguish three types of missingness that have implications for how missing data should be addressed (Little and Rubin, 2019). The first type, referred to as missing completely at random (MCAR), holds if the probability of missing values for an indicator is unrelated to the underlying values of that indicator. This would be true, for instance, in designs in which participants randomly receive different batteries of items. The second type, referred to missing at random (MAR), holds if the probability of missing values for an indicator is unrelated to the underlying values of that indicator conditional on any auxiliary variables included in the model. In the case of CFA, auxiliary variables are variables that predict missingness in any of the indicators of a latent variable. For instance, perhaps for a particular set of items women are less likely to respond than men. The measurement model may not include gender, but incorporating gender as an auxiliary variable could help ensure that the missing data are MAR. Enders (2010) provides a nice overview of how to incorporate auxiliary variables into CFA models, and some software for fitting CFAs (e.g., Mplus) automates these procedures. Finally, the third type, referred to as missing not at random (MNAR), holds if the probability of missing values for an indicator is related to the underlying values of that indicator. For example, if respondents who are more likely to rate a leader as unintelligent are more likely to skip the item and not provide a response, then the missing data for the *intelligent* item would be MNAR.

It is important for analysts to consider what types of missing data may be present because the DML estimator is based on the assumption that the data are either MCAR or MAR. If the data are in fact MNAR, then the resulting parameter estimates may be biased.[3] Other contemporary approaches to missing data, such as multiple imputation, rely on the same assumption that the data are MCAR or MAR.

To illustrate the use of the DML estimator and different types of missingness, we created two additional versions of the *intelligent* and *likeable* indicators with missing data. In the first version, we randomly discarded the values for 25% of the cases for each of the indicators. This approach to generating missing data creates an MCAR pattern of missingness. In the second version, we discarded values greater than 4 for each of the indicators (roughly the top 25% of the distribution for each indicator). In contrast to our first approach, this approach creates an MNAR pattern of missingness.

Table 3.4 reports estimates factor loadings from the standard ML estimator and from the DML estimator for both versions of missing data in the two indicators. We note that the analysis sample remains at 318 for both of the DML estimators. If instead an analyst used listwise deletion, then the analysis sample for the first version of missingness would be 172 and the analysis sample for the second version of missingness would be 160. Beginning with the MCAR setting, we see that even having discarded data on the *intelligent* and *likeable* indicators for 25% of the cases, we obtain quite similar estimates to the standard ML with all the data for the factor loadings. With our MNAR setting, however, we find substantially biased estimates of the factor loadings of the indicators with missing data, 0.79 as compared with 1.23 in the full-data analysis for *intelligent* and 0.79 as compared with 1.00 for *likeable*. In addition, the bias spreads to the other factor loadings as well, although it is not as significant in magnitude.

---

[3] In some cases, it may be possible to address MNAR data by modeling the missing data mechanism (see Enders, 2010, for a discussion of options).

Table 3.4: Comparison of maximum likelihood estimates with different forms of missing data for factor loadings from leadership measurement model.

	Standard ML	DML-MCAR	DML-MNAR
Leadership Ability			
Good leader	1 (—)	1 (—)	1 (—)
Intelligent	1.23 (0.09)	1.24 (0.10)	0.79 (0.06)
Influential	1.27 (0.11)	1.30 (0.12)	1.20 (0.11)
Strong	1.17 (0.09)	1.20 (0.09)	1.11 (0.08)
Competent	1.19 (0.09)	1.20 (0.09)	1.15 (0.08)
Likeability			
Nice	1 (—)	1 (—)	1 (—)
Pleasant	1.01 (0.03)	1.02 (0.03)	1.02 (0.03)
Likeable	1.00 (0.03)	0.99 (0.04)	0.79 (0.04)

Note: Unstandardized estimates with standard errors are in parentheses. Estimates wereobtained from Stata. ML = maximum likelihood; DML = direct maximum likelihood; MCAR = missing completely at random; MNAR = missing not at random.

DML estimators are widely available and often used to address missing data when fitting CFA measurement models. Before using a DML estimator, it is important for analysts to consider the potential underlying causes of missing data and, in particular, be mindful of the possibility of MNAR data. As we have illustrated, the DML estimator works well with MCAR (and also MAR) data but can produced substantially biased estimates with MNAR data.

3.2.4 Model-Implied Instrumental Variables

As mentioned above, ML estimators and variants that address nonnormally distributed variables and missing data are by far the most commonly used estimators for CFA. ML estimators, however, do suffer from some limitations. First, the ML estimators simultaneously estimate all the parameters in a given measurement model. In general, this is desirable from a statistical and practical perspective, as it represents an efficient use of the data and is convenient for researchers. Simultaneous estimation of parameters has one drawback in that if a component of the measurement model is misspecified (e.g., perhaps a cross-loading for an indicator is left out), then that error in specification can spread throughout the model and affect parameter estimates in other components of the model. This is akin to what we observed in the previous section with the MNAR data in one indicator leading to bias in the factor loadings for other indicators. Second, as discussed above, the ML estimator relies on an iterative numerical algorithm and can be subject to nonconvergence issues. Model-implied instrumental variable (MIIV) estimators are an alternative to ML estimators that address these limitations (Bollen, 1996).

To understand how MIIV estimators work, it helps to work through an example. We will focus on just the measurement model for latent likeability with the three indicators *nice*, *pleasant*, and *likeable*. Suppose we reference the indicators as x_1, x_2, and x_3, respectively, and use L for latent likeability. We can write the equations for this model as

$$x_{1i} = L_i + \delta_{1i} \tag{3.9}$$

$$x_{2i} = \alpha_2 + \lambda_2 L_i + \delta_{2i} \tag{3.10}$$

$$x_{3i} = \alpha_3 + \lambda_3 L_i + \delta_{3i}, \tag{3.11}$$

where the intercept is fixed at 0 and the factor loading at 1 in the equation for *nice*, the scaling indicator. We can rearrange Equation (3.9) to solve for the latent variable

$$L_i = x_{1i} - \delta_{1i} \tag{3.12}$$

and then substitute the equation for the latent variable into the remaining two equations

$$x_{2i} = \alpha_2 + \lambda_2(x_{1i} - \delta_{1i}) + \delta_{2i} \tag{3.13}$$

$$x_{3i} = \alpha_3 + \lambda_3(x_{1i} - \delta_{1i}) + \delta_{3i}. \tag{3.14}$$

We can reorganize Equations (3.13) and (3.14) to collect the error terms in each as

$$x_{2i} = \alpha_2 + \lambda_2 x_{1i} + (\delta_{2i} - \lambda_2\delta_{1i}) \tag{3.15}$$

$$x_{3i} = \alpha_3 + \lambda_3 x_{1i} + (\delta_{3i} - \lambda_3\delta_{1i}). \tag{3.16}$$

If we take a moment to look at Equations (3.15) and (3.16), we see that we have eliminated the latent variable likeability and instead have expressed the two nonscaling indicators as a function of the scaling indicator and a composite error term. At this point, we could fit a standard linear regression model for the first equation as both x_1 and x_2 (i.e., *nice* and *pleasant*) are observed variables, but the estimate for coefficient on x_1 (i.e., the factor loading λ_2) would be biased because x_1 is correlated with the composite error term due to the presence of δ_1. The same holds for the second equation.

Table 3.5: Comparison of maximum likelihood estimates model-implied instrumental variable estimates for factor loadings from likeability measurement model.

	Standard ML	MIIV-2SLS
Nice	1 (—)	1 (—)
Pleasant	1.02 (0.03)	1.02 (0.03)
Likeable	0.99 (0.03)	0.99 (0.03)

Note: Unstandardized estimates with standard errors are in parentheses. Estimates were obtained from Stata. ML = maximum likelihood; MIIV = model-implied instrumental variable; 2SLS = two-stage least squares.

One approach to addressing a correlation between a predictor and an error is to find an instrument or set of instruments for the endogenous variable and use one of the instrumental variable estimators, such as two-stage least squares (2SLS). For an instrument to be valid, it must be (1) uncorrelated with the error and (2) correlated with the endogenous predictor. Note that the first condition implies that the instrument is only correlated with the outcome due to its correlation with the endogenous predictor. If we return to our measurement model, we see that in the specified measurement model x_3 (*likeable*) meets these conditions. As specified, the indicator *likeable* is uncorrelated with both δ_1 and δ_2, so it is uncorrelated with the composite error term. In addition, *likeable* is correlated with *pleasant* due to the indicators' common dependence on latent likeability. A similar process reveals that *pleasant* is a valid instrument for *nice* in Equation (3.16). As such, the specified measurement model itself determines which, if any, of the indicators are suitable instruments. This is the reason we refer to the estimator as a "model-implied" instrumental variable estimator.

Once instruments have been identified for each equation, an analyst can proceed with an MIIV-2SLS estimator to obtain estimates of the factor loadings one equation at a time. Table 3.5 provides both standard ML estimates and MIIV-2SLS estimates for the factor loadings for the measurement model for latent likeability. We see that the estimated factor loadings and standard errors are identical to two decimal points for both estimators. This is due to having a just-identified model and only a single instrument for each equation. In general, the MIIV-2SLS estimates will not be identical to the ML estimators.

It is possible, of course, that the analyst has incorrectly specified the measurement model, in which case indicators that appear to be valid instruments based on the model may not in fact be valid. As we will discuss in the next chapter, the standard overidentification tests for valid instruments with 2SLS estimators work similarly to the overall model fit tests with ML estimators to detect evidence of model misspecification.

In summary, MIIV estimators provide a valuable alternative to ML estimators that address some of their limitations. MIIV estimators, however, come with some limitations of their own. First, it can be tedious to work through and reorganize the model equations to identify valid instruments. Programs are available in Stata (Bauldry, 2014) and R (Fisher et al., 2019) that assist analysts in identifying valid instruments for any given CFA and, with the R package, provide MIIV estimates. Second, MIIV estimators work well for obtaining estimates of factor loadings and intercepts, but not as well for obtaining estimates of error variances and covariances or means, variances, and covariances among latent variables.

3.3 Conclusion

In this chapter, we have moved from the specification of a CFA measurement model to an assessment of whether a proposed model is identified and various estimators for obtaining parameter estimates. In our next chapter, we turn to the question of how well a specified measurement model fits with the data.

3.4 Further Reading

There are a number of excellent sources that provide more detail about the various identification strategies and estimators we have discussed in this chapter. For an introduction to model identification in the broader SEM framework with specific chapters devoted to CFA, see Bollen (1989). Eliason (1993) provides a general introduction to ML estimators, Savalei (2014) has a nice discussion of how robust ML estimators work, and Enders (2010) covers ML estimators that account for missing data. For readers interested in learning more about MIIV estimators, see Bollen (1996, 2001). All these sources are situated in the broader SEM framework, and readers will benefit from having an understanding of how CFA fits into that framework (see Kline (2015) for an introduction and Bollen (1989) for a more rigorous account).

CHAPTER 4. MODEL EVALUATION AND RESPECIFICATION

An important and valuable feature of CFA lies in evaluating how well a proposed measurement model fits with a given dataset. In this context, a measurement model that has a good "fit" with the data is one that largely accounts for any patterns in the means, variances, and covariances among the observed indicators and generates "reasonable" parameter estimates (e.g., nonnegative estimates for variances or factor loadings that make sense in terms of their scale). In contrast, a measurement model that has a bad fit is unable to account for particular patterns in the means, variances, and covariates of observed indicators; implies patterns that do not exist; or generates unreasonable parameter estimates. As we will see, there is not a universally agreed-on single test of model fit that permits a sharp distinction between measurement models with good and bad fit, but rather we can think of model fit as lying along a continuum from poor to good with satisfactory or acceptable lying somewhere in between and dependent on specific research contexts. In addition, it is important to recognize that finding a measurement model with a good fit with the data does not preclude the possible existence of alternative measurement models with equally good or even better fits with the data.

For measurement models with sufficient degrees of freedom (i.e., overidentified models), the evaluation of model fit has two components. The first component involves testing and assessing overall model fit—the extent to which a measurement model reproduces the means, variances, and covariances among the observed indicators. There are a number of different types of metrics for analyzing overall model fit with different strengths and weaknesses. The second component involves examining various parameter estimates for more precise or localized evidence of concerns with model fit than can be gleaned from the overall model fit metrics. For just-identified models, there are no degrees of freedom available for tests of overall model fit, and only the second component of model fit can be assessed.

In addition to assessing how well a proposed measurement model fits with data, analysts may also be interested in the question of which among more than one competing measurement models provides the best fit with data. To the extent that the competing models are nested (discussed below), this question can be explored with nested model tests that operate similarly to the tests of overall model fit. In some cases, it is also possible to compare nonnested models using information criterion.

Table 4.1: Item prompts and responses for the Kessler Psychological Distress Scale indicators.

	All	...	None
During the past 30 days, how often did you feel			
nervous? (*nervous*)	1	...	5
restless or fidgety? (*restless*)	1	...	5
hopeless? (*hopeless*)	1	...	5
so sad or depressed that nothing could cheer you up? (*depressed*)	1	...	5
that everything was an effort? (*effort*)	1	...	5
down on yourself, no good, or worthless? (*worthless*)	1	...	5

Note: All = all of the time; None = none of the time.

In many cases, analysts will find that an initial proposed measurement model has a poor fit with the data. A theoretically or substantively informed model specification that does not fit with the data usually requires the analyst to rethink the theoretical or substantive justification of the model and/or the measurement properties of the indicators. This process of rethinking and exploring alternative model specifications may be theory- or substantively-driven, empirically-driven, or a mix of both. Analysts should recognize, however, that model respecification involves a shift from a *confirmatory* to a more *exploratory* mode of analysis and that results should be understood in that light.

For the first two sections of this chapter, we use the six-item version of the Kessler Psychological Distress Scale (K6; Kessler et al., 2002) included in the 2012 wave of the National Survey on Drug Use and Health (NSDUH) to demonstrate the concepts. Our examples draw on 5,431 young adult respondents (ages 26–34). Table 4.1 presents the indicators from this scale with responses ranging from 1 = *all of the time* to 5 = *none of the time*. For this chapter, we treat these indicators as continuous measures as is often done with Likert-type scale items involving five or more categories. Chapter 6 provides an overview of CFA with categorical indicators.

4.1 Model Evaluation

The two components of model evaluation involve assessing overall measurement model fit and examining parameter estimates for indications of poor model fit. To illustrate various options for each, we analyze a measurement

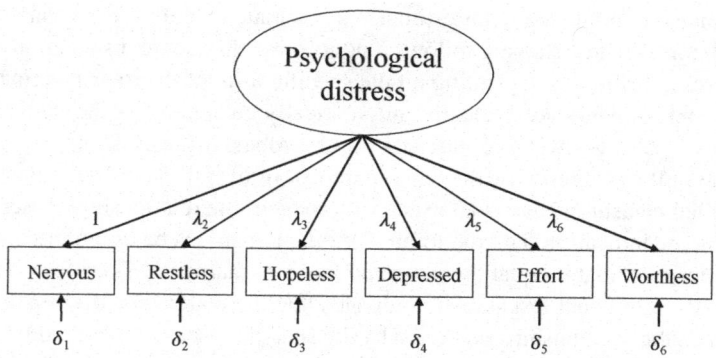

Figure 4.1: Measurement model of psychological distress using the Kessler Psychological Distress Scale indicators.

model based on the K6 indicators. The K6 was designed as a measure of psychological distress, and, therefore, we begin with a measurement model that specifies all six measures as indicators of a single latent variable representing psychological distress (see Figure 4.1). As ML estimators of the various types described in Chapter 3 are the most frequently used in practice, we rely on a the ML estimator here. In keeping with this, we focus our discussion of overall model fit on various tests and indices available with standard ML estimators.

4.1.1 Overall Model Fit

4.1.1.1 Chi-Square Test

The first approach to assessing the fit of a measurement model involves a chi-square test derived from the ML fitting function. The test statistic with a standard ML estimator is given by

$$T_{ML} = (N - 1)F_{ML}, \tag{4.1}$$

where N is the sample size and F_{ML} is the value of the ML fitting function. The test statistic follows a χ^2 distribution with degrees of freedom equal to the number of nonredundant elements in the mean vector and covariance matrix among the observed variables minus the number of parameters. In our case for the measurement model for psychological distress, we have 9 degrees of freedom for the chi-square test of overall model fit. As discussed in Chapter 3, it is possible to specify measurement models that have no degrees of freedom to test overall model fit. In such cases, analysts must rely on examining parameter estimates to evaluate model fit. For robust ML estimators, the test statistic follows a noncentral chi-square distribution that requires adjustments depending on the specific form of the robust estimator. These adjustments are typically automatically calculated in the statistical software packages when researchers request robust ML estimators.

The null hypothesis for the chi-square test of overall model fit is that the specified measurement model exactly reproduces the relationships observed among the indicators (i.e., the implied moment matrices based on the model parameter estimates equal the observed moment matrices). Therefore, a statistically significant test statistic indicates that there are some discrepancies between the relationships predicted by the specified measurement model and the relationships among the observed indicators. In other words, a statistically significant chi-square test statistic provides evidence of a lack of model fit. For our measurement model for psychological distress, we find a chi-square test statistic of 678.51 with 9 *df*, which has a *p* value of less than 0.001. This is our first indication that this measurement model has a poor fit with the data.

Although the chi-square test statistic is theoretically sound, it has a couple of characteristics that have led psychometricians to develop alternative methods of assessing model fit and for researchers rarely to rely on it alone. First, the chi-square test is based on the strict hypothesis that the implied moment matrices perfectly reproduce the observed moment matrices. In practice, this is often an unrealistic expectation for a statistical model, and we would prefer not to discard models that are sufficiently good approximations to reality to prove useful. Second, the chi-square test statistic increases in power with increasing sample size. As such, at large sample sizes, the test statistic has the ability to detect relatively minor deviations from perfect fit. This has the perverse effect of encouraging the use of small sample sizes to avoid the power to detect minor measurement model misspecifications.

4.1.1.2 Model Fit Indices

Numerous alternatives to the chi-square test for overall model fit have been developed. In this section, we present four of the most frequently reported indices: (1) standardized root mean square residual (SRMR), (2) the root mean square error of approximation (RMSEA), (3) the Tucker–Lewis index (TLI), and (4) the comparative fit index (CFI).

The first model fit index, the SRMR, is similar to the chi-square test statistic in that it is based on the idea of perfect fit (Bentler, 1995). The SRMR is calculated as the average difference (i.e., residual) between the standardized implied moment matrices and the observed moment matrices. The index takes values between 0 and 1 with values closer to 0 (i.e., smaller residuals) indicative of better fit. Simulation studies suggest that values below 0.08 suggest reasonable fit for the specified measurement model (Hu and Bentler, 1999). We find an SRMR of 0.04 for our measurement model for psychological distress, which indicates reasonable fit.

The second model fit index, the RMSEA, relaxes the idea of perfect fit and instead considers whether a measurement model has an approximate fit (Steiger and Lind, 1980). This shift is accomplished by estimating a noncentrality parameter that derives from a noncentral chi-square distribution that holds when a measurement model does not have a perfect fit. The noncentrality parameter can be estimated as the difference between the chi-square test statistic and the degrees of freedom divided by the sample size minus 1 and the RMSEA is given by

$$\text{RMSEA} = \sqrt{\frac{\max(T_{ML} - df, 0)}{df(N-1)}}. \tag{4.2}$$

The RMSEA has a minimum of 0 but no upper limit. Values below 0.06 are considered indicative of good fit and values between 0.06 and 0.10 are considered indicative of adequate fit. With standard ML estimators, the use of the noncentral chi-square distribution allows for the calculation of confidence intervals and a test for the null hypothesis that the true value of the RMSEA is below 0.05. RMSEA, however, is known to be overly conservative in sample sizes less than 200 (Curran et al., 2003). We find an RMSEA of 0.12 for our measurement model for psychological distress. This is another indicator of poor fit.

The remaining two model fit indices that we discuss, the TLI and CFI, are both derived from comparing the fit of the specified model against the fit of a more restricted baseline model (Bentler, 1990; Tucker and Lewis, 1973). For these indices, the baseline model, sometimes referred to as a "null" or "independence" model, specifies that there are no relationships among all the observed indicators. If we let T_m and df_m refer to the chi-square test statistic and degrees of freedom for the specified model and T_b and df_b refer to the chi-square test statistics and degrees of freedom for the baseline model, then we can express the TLI and CFI as

$$\text{TLI} = \frac{T_b/df_b - T_m/df_m}{T_b/df_b - 1} \tag{4.3}$$

$$\text{CFI} = \frac{\max(T_b - df_b, 0) - \max(T_m - df_m, 0)}{\max(T_b - df_b, 0)}. \tag{4.4}$$

Both indices have lower limits of 0 and upper limits close to 1 or in some versions fixed to 1. Simulation studies suggest that values above 0.95 indicate good fit (Hu and Bentler, 1999). For our measurement model for psychological distress, we find a TLI of 0.93 and a CFI of 0.96, providing mixed evidence of model fit.

4.1.1.3 Information Criterion

The and Bayesian information criterion (BIC) are criteria for model selection among a set of proposed models (Akaike, 1973; Raftery, 1995). As discussed in the next section, in the context of CFA they can be used to evaluate the relative fit of one or more alternative measurement models. One particular form of the BIC, however, can also be useful for evaluating the overall fit of a given model. This form of the BIC, sometimes referred to as the Schwarz-modified BIC (SBIC) or Schwarz information criterion (SIC), can be calculated from the chi-square test statistic as

$$\text{SBIC} = T_{ML} - df * \ln(N). \tag{4.5}$$

The utility of this form of the BIC derives from the ability to compare it with the SBIC of a "saturated model" in which all indicators are allowed to be correlated and has 0 degrees of freedom. The SBIC for a saturated model is 0, which facilitates comparisons with the specified model. In this form (as in other forms of the AIC and BIC), lower values of the index support the given model. Thus, a negative SBIC provides support for the specified model relative to the saturated model, while a positive SBIC supports the saturated model over the specified model. In this context, a positive SBIC provides evidence that the specified model has poor fit with the data. For our measurement model for psychological distress, we find an SBIC of 601.11.

4.1.1.4 Summary

As is evident from our discussion in this section, we have multiple tests and indices to aid in our assessment of the overall fit of a measurement model. In practice, we recommend interpreting and reporting the chi-square test statistic and a number of additional indices. In some cases, all the various tests and indices will point toward the same conclusion with regard to overall model fit. In other cases, however, the tests and indices will provide conflicting information regarding model fit, as is true for our measurement model for psychological distress. In such situations, analysts will need to rely on their judgment that takes into account the context for the research to determine whether model fit is adequate or whether alternative models should be considered. For our example, given that we are using indicators that were designed to be measures of psychological distress, we would expect better fit with the data than we observe.

The model fit statistics presented above are all based on a standard ML estimator. Various robust ML estimators and other estimators for CFAs (e.g., MIIV estimators) have more limited options to assess overall measurement model fit. Most other estimators have at least an analogue to the overall chi-square test, such as an adjusted chi-square test available with robust ML estimators or an overidentification test with MIIV estimators, but the other indices to assess overall measurement model fit are not necessarily available or routinely reported in statistical software.

4.1.2 Item-Level Model Fit

The second component of model evaluation involves examining the parameter estimates for potential indications of poor model fit. This can take two forms. The first form involves checking for inadmissible parameter estimates. As described in Chapter 3, inadmissible estimates are those that take theoretically impossible values, such as a negative estimate for a variance or a correlation that lies outside the bounds of -1 and 1. Such estimates can be a sign of measurement model misspecification.

The second form involves examining the interpretability and size of the parameter estimates. This may include examining the direction and size of the various parameter estimates. Among the factor loadings, analysts should expect to see that the signs are in the expected directions. For instance, the indicators for psychological distress are all scored such that the higher values signify greater psychological distress. If the factor loading for one of the indicators turned out to be negative, while the others were positive, this could suggest a problem with the measurement model (or, alternatively, an error in data preparation). Beyond the direction of the factor loadings, analysts should also expect to see that any latent variables have clear relationships with the indicators. This can be assessed by looking at either the unstandardized or standardized factor loadings relative to their standard errors. Indicators with factor loadings close to 0 may be candidates for trimming or an indication that a latent variable is not capturing the expected concept. Similarly, factor loadings of markedly different sizes (either unstandardized with indicators in roughly the same scale or standardized) may be an indication that a given latent variable is not in fact unidimensional and should be split into two latent variables. Although less common, in some research contexts, analysts may also be interested in examining the estimates of intercepts for direction and size.

Table 4.2: Selected parameter estimates for the psychological distress measurement model.

	Factor Loading		R^2
	Unstandardized	Standardized	
nervous	1 (—)	0.66	0.43
restless	1.02 (0.02)	0.65	0.42
hopeless	1.18 (0.03)	0.83	0.68
depressed	1.13 (0.03)	0.84	0.70
effort	1.12 (0.03)	0.62	0.38
worthless	1.19 (0.03)	0.83	0.69

Note: Maximum likelihood estimates with standard errors in parentheses. National Survey on Drug Use and Health persons aged 26 to 34 years; $N = 5,431$. Parameter estimates were obtained from Stata.

In addition to checking for negative error variances or any correlations between errors outside the range of -1 to 1, it can be useful to check for relatively high error variances across similarly scaled indicators. Such high error variances also reveal themselves as low indicators R^2s. As mentioned in Chapter 3, the indicator R^2s reflect the amount of variance in an indicator explained by a latent variable (or multiple latent variables if the indicator loads on more than one). This can be interpreted as a measure of the reliability of the indicator, which can point toward indicators to consider dropping from the model or as evidence that a latent variable is not capturing the expected concept. For instance, suppose a researcher pilot tests 10 indicators designed to measure attachment to a community organization. If three of the indicators have much lower R^2s than the others (e.g., R^2s around 0.2, while the others have R^2s around 0.8), then the researcher might suspect that the three indicators are not good measures of attachment or alternatively that attachment to a community organization is a multidimensional concept and the three indicators are capturing a different dimension than the other seven indicators.

Finally, analysts should check that estimates of variances and any covariances among latent variables are reasonable in magnitude and any covariances are in expected directions.

Table 4.2 reports selected parameter estimates for our measurement model for psychological distress. Of note, the R^2 for *effort* is low relative to others, as are R^2s for *nervous* and *restless*. There also appears to be a pattern in the standardized loadings with *hopeless*, *depressed*, and *worthless* all loading on the latent variable psychological distress at a slightly higher level than the other three indicators.

In summary, the combination of the overall model fit statistics and an exploration of the parameter estimates suggest that our measurement model for psychological distress has an inadequate fit with the data. At this point, we would need to consider alternative models. In some cases, researchers may have alternative models in mind and can proceed with evaluating the fit of the alternatives. In other cases, researchers may need to develop alternative models based on a reconsideration of the theoretical or substantive basis for the initial measurement model that may be informed by some of the indicators of poor fit from the initial model. If, however, the fit statistics for the psychological distress measurement model had indicated a reasonable fit with the data, then we would proceed with interpreting the parameter estimates. It is important to recognize, however, that finding evidence that a given measurement model has a good fit with the data does not preclude the possibility that alternative measurement models exist with equal or even better fit with the data.

4.2 Comparing Models

In some research contexts, it can be useful not only to test a measurement model's overall fit with the data but also the relative fit of one measurement model versus another. For instance, in our example with psychological distress, we might posit that the errors for *nervous* and *restless* share a common source of variance other than latent psychological distress, and we might be interested in testing whether a model that incorporates a covariance between these two errors has a better fit with the data than the initial model. With CFA, we have two approaches for conducting such tests: (1) chi-square difference tests for nested models and (2) assessments of changes in information criterion.

Before turning to the two approaches, it is important to understand how *nested models* are defined. Suppose we start with an initial measurement model that we refer to as an *unrestricted model*. This model has a set of parameters (factoring loadings, intercepts, error variances and covariances, and latent variable variances and covariances) associated with it. Any alternative measurement model that can be represented with a strict subset of the parameters of the unrestricted model is nested within the unrestricted model. We can refer to such a model as a *restricted* model in the sense that it imposes some restrictions on the parameters of the unrestricted model. For instance, for our psychological distress model, suppose we considered the alternative measurement model mentioned above that added a correlation between the errors for the first two indicators. In this case, the alternative model would represent an unrestricted model and the initial model would be a restricted form of the alternative model in which the covariance between the two errors is fixed to 0. It is also the case that constrained parameters yield nested models; model constraints are less restrictive than fixed parameters, but they are still more restrictive than freely estimated parameters.

4.2.1 Chi-Square Difference Test

With a standard ML estimator, it is straightforward to test the fit of two nested measurement models by conducting a chi-square difference test. The null hypothesis for this test is that the restricted measurement model fits the data as well as the unrestricted measurement model. Since the restricted measurement model involves estimating fewer parameters, it is preferred as long as it fits the data as well as the unrestricted measurement model. If we let T_R and df_R refer to the chi-square test statistic and associated degrees of freedom for the restricted measurement model, respectively, and T_U and df_U refer to the chi-square test statistic and degrees of freedom for the unrestricted measurement model, respectively, then the chi-square difference test is constructed as

$$T_d = T_R - T_U \tag{4.6}$$

$$df_d = df_R - df_U. \tag{4.7}$$

A statistically significant result supports the unrestricted measurement model, while a nonsignificant result supports the restricted measurement model. Chi-square difference tests are also possible with robust ML estimators, but adjustments need to be made to both the chi-square test statistics and the degrees of freedom for the restricted and unrestricted measurement models (Satorra and Bentler, 2010).

Consider two models: one, we'll refer to as Model 1, represented by Figure 4.1, and the other, Model 2, represented in Figure 4.2 in which the error covariance between *nervous* and *restless* is freely estimated. This costs 1 degree of freedom, and Model 1 is nested within Model 2. Both models contain the same set of observed variables and are estimated on the same sample. Model 1, our restricted model, has a chi-square test statistic of 678.51 with 9 degrees of freedom, while Model 2, our unrestricted model, has a chi-square test statistic of 82.78 with 8 degrees of freedom. A chi-square difference test

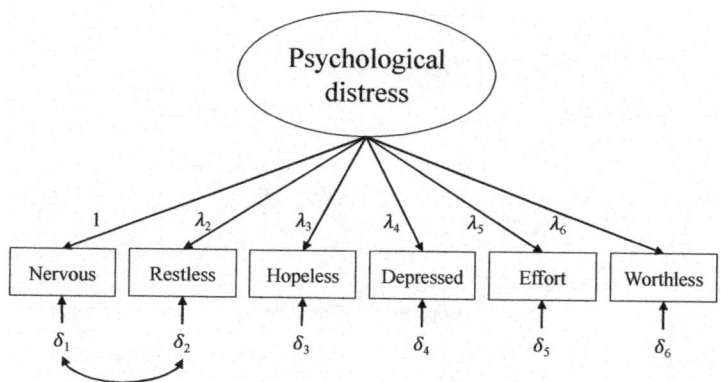

Figure 4.2: Measurement model of psychological distress using the Kessler Psychological Distress Scale indicators with constraint on covariance between *nervous* and *restless* relaxed.

is easily calculated as

$$T_d = 678.51 - 82.78 \tag{4.8}$$
$$df_d = 9 - 8. \tag{4.9}$$

This results in a T_d of 595.73 and a df_d of 1, which has a p value less than 0.001. This result of the test indicates that allowing for the covariance between the errors for *nervous* and *restless* results in a model with a substantially (and statistically significantly) better fit with the data and thus the unrestricted model (Model 2) is preferred over the restricted model (Model 1).

4.2.2 Information Criterion

Chi-square difference tests are quite valuable in testing nested measurement models, but in some contexts, analysts may be interested in selecting among nonnested measurement models. For instance, it is often the case that models with different latent variables are not nested. In our example with psychological distress, we might consider an alternative model that splits psychological distress into two latent variables, one that captures anxiety and another that captures depression. In this case, the two models are not nested, and we

60

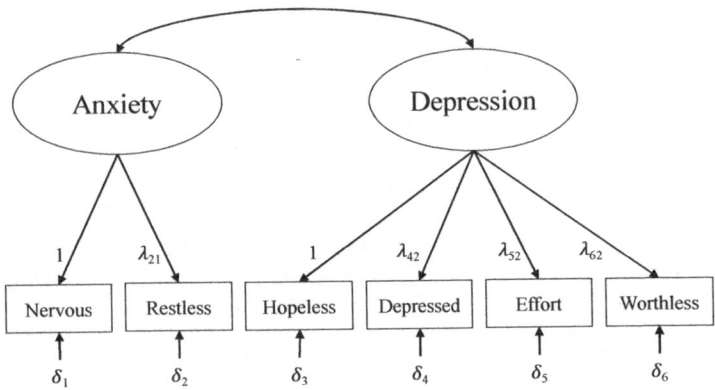

Figure 4.3: Measurement model of anxiety and depression as two dimensions of psychological distress using the Kessler Psychological Distress Scale indicators.

would need an alternative to a chi-square difference test in selecting between the two.[1]

As long as the alternative measurement models have the same sample and same observed variables, information criterion, such as the AIC and BIC, can be used as a basis for selecting among alternative models. As noted above, in general smaller values of an information criterion lend support to a given model relative to an alternative model. For the BIC (or SBIC) a widely used set of thresholds is that a difference between 0 and 2 indicates "weak" evidence, 2 and 6 indicates "positive" evidence, 6 and 10 indicates "strong" evidence, and greater than 10 indicates "very strong" evidence in support of the model with the smaller BIC (Raftery, 1995).

Returning to our psychological distress example, rather than permitting a correlation between the errors for *nervous* and *restless*, we might instead posit that these two items represent measures of anxiety while the remaining four items represent measures of depression. In other words, our measurement model should reflect two dimensions of psychological distress, anxiety and depression, that are correlated rather than a single dimension. Figure 4.3

[1]One can make the case that these models are nested in the sense that the measurement model with one latent variable is a restricted version of the model with two latent variables in which the correlation between the two latent variables is fixed to 1. This is true, but it requires testing a parameter restriction on the boundary of permissible values (correlations cannot be greater than 1) that violates an assumption of the chi-square difference test.

Table 4.3: Selected model fit statistics for two models based on Kessler Psychological Distress Scale indicators of psychological distress.

	One-Dimension	Two-Dimension
Chi-square	678.51	82.78
df	9	8
p Value	<0.001	<0.001
SRMR	0.04	0.01
RMSEA	0.12	0.04
TLI	0.93	0.99
CFI	0.96	1.00
SBIC	601.11	−13.98

Note: National Survey on Drug Use and Health persons aged 26 to 34 years; $N = 5{,}431$. Model fit statistics were obtained from Stata. SRMR = standardized root mean square residual; RMSEA = root mean square error of approximation; TLI = Tucker–Lewis index; CFI = comparative fit index; SBIC = Schwarz-modified Bayesian information criterion.

illustrates this model as an alternative to the unidimensional model for psychological distress depicted in Figure 4.1.

As noted above, these two models are not nested such that a chi-square difference test is valid, but we can evaluate their overall fit with the data and the difference in SBICs across the two models. Table 4.3 reports selected model fit statistics for both models. We see that the model allowing for separate dimensions for anxiety and depression has a reasonable fit with the data and is clearly preferred based on a difference in SBICs of 615.09. As such, we have strong support that the two-dimensional model fits the data better than the one-dimensional model, although there is still room for improvement in the overall model fit for the two-dimensional model.

4.2.3 Chi-Square Equivalent Models

When specifying alternative measurement models, it is possible to consider two different measurement models that result in identical chi-square test statistics and model fit indices. This occurs when a given model structure implies the same structure among the observed means, variances, and covariances as an alternative model structure.[2] In such cases, both measurement models fit the data equally well and it is not possible to adjudicate between them based on statistical criteria.

In fact, our consideration of the two alternative models for psychological distress illustrate a well-known example of this. The model permitting a correlation between the errors of the first two indicators and the model specifying a separate latent variable measured by the first two indicators represent two different ways of accounting for shared variance in the first two indicators that is not accounted for by a single latent variable for psychological distress. As such, the models imply the same pattern of relationships among the observed means, variances, and covariances and, therefore, the overall model fit statistics for the two models are identical. In this case, we are unable to decide between the two empirically and instead must rely on theoretical or substantive considerations to determine which should be reported.

More generally, the existence of chi-square equivalent models is a reminder of a fundamental asymmetry in our evaluation of how well measurement models fit with data. We can interpret evidence of a lack of fit that our measurement model is misspecified in potentially important ways. But we need to be cautious in interpreting evidence of adequate or good fit that our measurement model is *correct* as there are always alternative models that may fit the data equally well or even better.

4.3 Model Respecification

In many cases, researchers may find that an initial measurement model or set of alternative measurement models has an inadequate fit with data. In these contexts, it is often of interest to explore potential improvements to the measurement model specification that could result in a better fit with the data. There are two approaches to doing so. First, researchers may take another look at the indicators and theory or substantive knowledge underlying the initial measurement model specification and consider whether there are reasonable alternatives. For instance, as illustrated with our psychological distress example, researchers could consider whether a latent variable assumed to have one dimension actually has multiple dimensions, whether the wording of questions might suggest correlations among selected indicator errors, or whether some measures do not perform as expected as indicators of a

---

[2]This is a special case of the broader concept of "observational equivalence" that refers to the situation when two or more models generate the same probability distribution of observed data.

latent variable. These and many other possibilities can form the basis for respecifying the initial measurement model.

The second approach is empirical in nature and typically involves exploring changes to the measurement model based on the improvement in fit that would come from freeing one parameter at a time. This approach relies on *modification indices* that can be requested following fitting a measurement model in most statistical software packages. Modification indices are calculated for each parameter that is fixed or constrained in a model. This includes parameters that are not explicitly fixed to a value by a researcher but rather implicitly fixed to 0 by the structure of the model (e.g., error covariances or indicator cross-loadings). The modification index for a given parameter reflects an approximation of how much the overall chi-square test statistic would be reduced if that parameter were instead free to be estimated. With this approach, parameters with relatively high modification index values are candidates for alterations to the original model. For instance, in our psychological distress example, the modification index for the covariance between the errors for *nervous* and *restless* is 597.7, relative to a χ^2 statistic of 678.51. Given that it would only cost 1 degree of freedom to free this parameter to be estimated, this would lead to a very substantial improvement in model fit.

Using model fit indices to guide the respecification of measurement models comes with a number of cautions. First, the approach is atheoretical and can sometimes lead to nonsensical measurement models that involve implausible or uninterpretable parameter estimates. Any use of modification indices should at a minimum be guided by substantive and theoretical knowledge to avoid this pitfall. Second, modification indices are limited in the sense that they only consider freeing fixed or constrained parameters in a given model structure and do not offer guidance toward alternative model structures. For instance, the modification indices would not be able to point to a two-dimensional measurement model for the indicators of psychological distress based on the initial one-dimensional measurement model. Third, modification indices are also limited in the sense that they only consider freeing one parameter at a time. This suggests the possibility of an iterative search in which a series of models are fit and modification indices are requested with each successive model freeing a parameter until the measurement model has a reasonable fit with the data or there are no longer any degrees of freedom remaining. Such an iterative approach, however, has been demonstrated in simulation studies to perform poorly and to rarely converge on the population measurement model (MacCallum et al., 1992).

Given these cautions, we recommend either relying on substantive or theoretical knowledge or a combination of substantive or theoretical knowledge and empirical approaches informed by that knowledge to guide model

Table 4.4: Item prompts and responses for self-efficacy indicators.

	SA	...	SD
"How much do you agree or disagree with the following statements?"			
There is little I can do to change the important things in my life. (*change*)	1	...	5
Other people determine most of what I can and cannot do. (*determine*)	1	...	5
There are many things that interfere with what I want to do. (*interfere*)	1	...	5
I have little control over the things that happen to me. (*control*)	1	...	5
There is really no way I can solve the problems I have. (*solve*)	1	...	5

Note: SA = strongly agree; SD = strongly disagree.

respecification. It is unwise to rely on empirical approaches and, in particular, modification indices alone to guide this process. Furthermore, as emphasized in the introduction, researchers should realize and acknowledge that once they begin exploring model respecifications, CFA takes on a more "exploratory" rather than "confirmatory" character.

4.3.1 Extended Example

To illustrate the process and some of the pitfalls of model respecification using modification indices, we draw on an example of a measurement model for self-efficacy. The data for this example come from the public-use version of Wave 4 of the National Longitudinal Study of Adolescent to Adult Health (Add Health; Harris et al. 2009) and includes five measures of self-efficacy (see Table 4.4). The responses for each measure range from 1 "strongly agree" to 5 "strongly disagree." As with our psychological distress example, we treat these Likert-type scale items as continuous measures and use the ML estimator (we cover measurement models with categorical indicators in Chapter 6).

We begin with a baseline measurement model with a latent variable for self-efficacy measured by all five indicators (see Figure 4.4). This model has a mediocre fit with the data. On the one hand, the RMSEA of 0.07, the TLI of 0.97, and the CFI of 0.98 point towards a reasonable fit with the data. On the other hand, the chi-square test for overall model fit is statistically significant and the SBIC is 69.95, which are both indicative of poor fit. Given that these

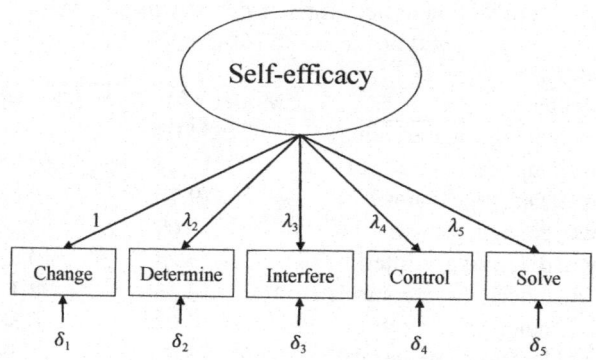

Figure 4.4: Baseline measurement model for self-efficacy.

indicators are derived from a validated scale, we might expect better model fit.

Suppose we now turn to the modification indices to see what parameters, if any, if freed would result in a substantial improvement in fit. To illustrate the potential dangers with this approach, we first randomly split the sample in half to follow modification indices on half of the sample and then check how well the resulting model fits in the remainder of the sample.[3]

We first refit our measurement model on the subsample and, as expected given the random selection of our subsample, we see similar indications of mediocre or inadequate fit. Table 4.5 reports the modification indices for our measurement model in the column labeled "Model 1 MI." In this case, all the parameters refer to covariances among the errors for different indicators. We see, for instance, that allowing for a covariance between the errors for *change* and *determine* would result in approximately a 34-point reduction in the chi-square test statistic from our base model. The chi-square test statistic for our initial model is 59.47, so allowing for this covariance would lead to a substantial improvement in model fit.

Indeed, refitting the measurement model allowing for a covariance between the errors for *change* and *determine* results in a model with much better fit to the data. This change yields a χ^2 of 26.81, at the cost of one degree of freedom. Suppose, however, we continue to explore additional respecifications with the modification indices for our respecified model. Table 4.5 reports the

---

[3]Cross-validation along these lines can be a useful approach in general with CFA when it is feasible (Browne and Cudeck, 1993).

Table 4.5: Modification indices for self-efficacy measurement models based on a random subsample.

Parameter	Model 1 MI	Model 2 MI
cov(e.change, e.determine)	34.04	—
cov(e.change, e.interfere)	18.27	10.95
cov(e.change, e.control)	9.70	0.64
cov(e.change, e.solve)	0.09	9.72
cov(e.determine, e.interfere)	3.24	10.83
cov(e.determine, e.control)	13.53	2.14
cov(e.determine, e.solve)	10.03	0.88
cov(e.interfere, e.control)	13.62	7.39
cov(e.interfere, e.solve)	2.21	8.23
cov(e.control, e.solve)	15.65	0.04
model chi-square	59.47	26.81

Note: Random subsample of Wave 4 Add Health; $N = 2,541$. Modification indices obtained from Stata. MI = modification index.

modification indices for our respecified measurement model in the column labeled "Model 2 MI."

With the respecified model, the covariance between the errors for *change* and *interfere* has the highest modification index, though the covariance in the errors for *determine* and *interfere* is close behind. Allowing for a correlation between the errors for *change* and *interfere* results in another improvement in model fit. The chi-square test statistic at 15.8 with 3 degrees of freedom remains statistically significant, but the SBIC is −7.72.

Figure 4.5 illustrates the alternative model for self-efficacy based on two iterations of the modification indices. We now turn to our hold-out sample to see how well this measurement model fits in the remaining subsample of cases. Despite finding a good fit with the data in our first subsample, in the hold-out sample, we find evidence of poor fit with an overall chi-square test statistic of 28.29 with 3 degrees of freedom and an SBIC of 4.76.

The purpose of our extended example is to illustrate the pitfalls of relying solely on an empirical approach to model respecification. Such an approach runs a high risk of capitalizing on idiosyncratic aspects of samples in developing a measurement model that has a low likelihood of replicating in other samples.

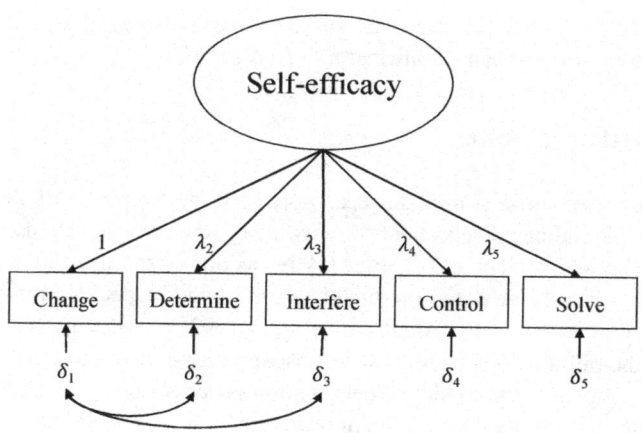

Figure 4.5: Alternative model for self-efficacy based on modification indices.

4.4 Conclusion

This chapter introduced the concepts of measurement model fitting, selection, and respecification alongside two empirical examples of measurement models. The ability to evaluate the overall fit of a model or the relative fit of alternative models is a powerful feature of CFA. It is important to keep in mind, however, that finding a measurement model that has an adequate or good fit with a given dataset does not preclude the possibility that alternative measurement models could fit the data equally or even better than the specified measurement model. In addition, some measurement models do not have sufficient degrees of freedom for overall model fit tests or are chi-square equivalent with alternative measurement models.

In this chapter, we also emphasized that in many cases initial specifications of measurement models will result in a poor fit with a given dataset, and researchers will need to consider revisions. These revisions should be guided by theory or substantive knowledge rather than solely data driven. We illustrated the pitfalls of data-driven approaches based on modification indices. In many contexts, researchers will not have the luxury of testing their measurement models on additional samples, and thus any aspects of a measurement model that capitalize on idiosyncratic variation in a given sample will go undetected. Even model respecification guided by theory or substantive knowledge, however, should be acknowledged as more in an "exploratory" rather than "confirmatory" framework.

4.5 Further Reading

Numerous simulation studies have explored the performance of model fit statistics and indices in a broader SEM context. Readers interested in learning more should see Hu and Bentler (1999) as one of the most highly cited of these studies. More detail on model selection and respecification may be found in West et al. (2012). MacCallum et al. (1992) provides a classic treatment of the pitfalls associated with empirically based model respecification. For an exemplar of the model respecification process, see Roos (2014), and for another with model respecification drawing from a much larger pool of novel items, see Manglos-Weber et al. (2016).

CHAPTER 5. MEASUREMENT INVARIANCE

In this chapter, we introduce the concept of measurement invariance and two approaches to investigate it. In conducting CFA on samples representing heterogeneous populations a natural question concerns the equivalence of measurement models across different subpopulations. Measurement models that have the same properties, as detailed below, in different groups are said to exhibit measurement invariance. For instance, we might investigate whether the measurement model for leadership discussed in Chapter 2 are equivalent for women and men or for people in different types of organizations. If so, then comparisons of mean levels of leadership between women and men or between members of different types of organizations are valid. More generally, there are several forms of equivalence that can be tested and different implications of establishing invariance for each.

An assessment of measurement invariance arises with a number of uses of CFA. Analysts using CFA for scale construction may examine measurement invariance for a set of indicators and discard those that do not exhibit invariance. In this context, including indicators that perform differently with respect to a latent variable across different subpopulations in a scale has the potential to introduce bias when using the scale in subsequent analyses. Suppose, for example, that an indicator for an anxiety scale has a stronger relationship with latent anxiety for older adults than for younger adults. If this indicator is included in the anxiety scale, then relationships between covariates and the anxiety scale may in part reflect age. Analysts may also be interested in using CFA to examine differences in means, variances, or covariances of latent variables across groups or over time. To do so, it is important to first establish that the measurement of the latent variables is invariant (or, as discussed below, at least partially invariant) across groups or over time.

There are two classic approaches to investigating measurement invariance with different strengths and limitations. The first approach involves multiple-groups CFA in which the structure of the measurement model itself and all the parameters of a measurement model are allowed to vary across a typically small number of groups. This approach allows for a complete assessment of the equivalence of measurement models across different groups, but it comes with the limitation that it is difficult to implement as the number of groups increases or when subpopulations are defined by continuous variables. The second approach involves incorporating covariates representing groups into CFA through the specification of a form of a MIMIC model.

Table 5.1: Prompt and indicators of beliefs about democracy.

Prompt: Many things are desirable, but not all of them are essential characteristics of democracy. Please tell me for each of the following things how essential you think it is as a characteristic of democracy. Use this scale where 1 means "not at all an essential characteristic of democracy" and 10 means it definitely is "an essential characteristic of democracy."	Not Essential		Essential
People choose their leaders in free elections. (*free*)	1	...	10
Civil rights protect people from state oppression. (*civil*)	1	...	10
Women have the same rights as men. (*women*)	1	...	10
Religious authorities ultimately interpret the laws. (*religious*)	1	...	10

This approach primarily focuses on assessing measurement invariance in the intercepts of the indicators. As such, it does not provide as complete an assessment of measurement invariance as the multiple-groups approach, but it is more flexible with respect to the types of covariates that can be explored and is generally less complex to implement.

To illustrate both approaches, we explore a measurement model for beliefs about democracy. The CFA includes four items from Wave 6 of the World Values Survey (Inglehart et al., 2014) that capture beliefs concerning rights and free elections for respondents in two countries, Germany and Spain (see Table 5.1). In these data, we have 1,885 respondents from Germany and 1,004 respondents from Spain. All the indicators are skewed, and, therefore, we use a robust ML estimator as described in Chapter 3. Our primary interest with this example lies in investigating the equivalence of measurement models for beliefs about democracy across respondents in Germany and Spain.

5.1 Multiple-Groups CFA

To assess measurement invariance using multiple-groups CFA, we begin by specifying the same measurement model and evaluating the fit for each group. Figure 5.1 illustrates our measurement model for democracy for our two groups of respondents from Germany and Spain. In both groups, we have latent democracy measured by four indicators, *free*, *civil*, *women*, and *religious*, with *free* specified as the scaling indicator. In the figure, we

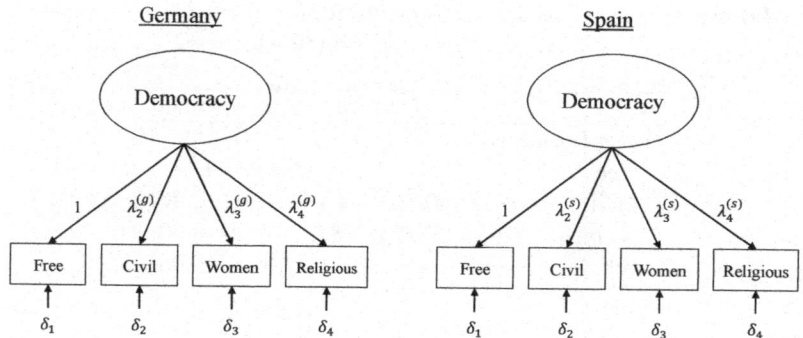

Figure 5.1: Measurement model for democracy in Germany and Spain.

illustrate that the factor loadings for *civil*, *women*, and *religious* may differ across groups by including a superscript *g* or *s* representing Germany and Spain, respectively. Although not depicted in the figure, the remaining model parameters—the intercepts, error variances, latent democracy mean, and latent democracy variance—may also differ across groups.

Before turning to the parameter estimates, we note that the conventional measurement model identification strategy for multiple-groups CFA differs from the strategy presented in Chapter 3. In particular, rather than fixing the intercept for the scaling indicator to 0, instead the mean of the latent variable is fixed to 0. This ultimately facilitates cross-group comparisons of the mean of the latent variable with one group serving as the referent group with a latent mean of 0 and the latent means of the other groups expressed as deviations from the referent group.

With four indicators, we have 2 degrees of freedom to test the overall fit of the model. The chi-square tests for overall model fit are nonsignificant for both countries, and the other indices of model fit all point to a measurement model that has a reasonable fit with the data among both German and Spanish respondents. We can also evaluate components of model fit for each group. Table 5.2 provides selected parameter estimates for measurement model for each group. We see that, for both Germany and Spain, the factor loadings indicate clear relationships between latent democracy and each of the nonscaling indicators. We also see that the estimates of intercepts for each indicator near the top (or bottom) of the range of values that reflect the overall high (or low) ratings that respondents in both countries give for how essential each indicator is as a characteristic of democracy. Finally, we note similar estimates of indicator error variances, although the

Table 5.2: Selected parameter estimates for democracy measurement model for Germany and Spain.

	Germany	Spain
Factor Loadings		
free	1 (—)	1 (—)
civil	0.86 (0.08)	1.02 (0.06)
women	0.79 (0.08)	1.00 (0.06)
religious	−0.77 (0.08)	−0.61 (0.06)
Intercepts		
free	9.28 (0.04)	8.63 (0.06)
civil	8.20 (0.05)	8.07 (0.06)
women	9.33 (0.04)	8.83 (0.06)
religious	2.02 (0.05)	3.42 (0.09)
Error Variances		
free	1.48 (0.22)	1.32 (0.15)
civil	4.16 (0.24)	1.92 (0.19)
women	1.89 (0.18)	1.08 (0.16)
religious	3.30 (0.26)	7.03 (0.37)

Note: Unstandardized estimates with standard errors in parentheses. Estimates were obtained from Mplus using MLR estimator.

estimates for the error variances for *civil* among German respondents and for *religious* among German and Spanish respondents are notably higher than the other estimates.

Having established and evaluated a baseline measurement model for each country, the next step is to engage in a sequence of tests of nested models that impose increasing constraints on the parameters across the groups (Meredith, 1993). Our first test is a simultaneous test of whether the same specification of the measurement model fits for each of the groups. This tests what is commonly referred to as *configural invariance* (or *equal form*). One important consideration for this test (and subsequent tests) is the distribution of cases across groups. Groups that represent a higher proportion of the overall sample contribute more to the chi-square test than groups that represent smaller proportions of the sample. For instance, in our data about 65% of the respondents are from Germany, so the test of configural invariance will place a bit more weight on the fit among German respondents than among Span-

Table 5.3: Selected model fit statistics for measurement invariance tests.

	χ^2	df	p Value	SBIC	RMSEA
Configural invariance	1.84	4	0.765	-30.03	0.00
Metric invariance	14.21	7	0.048	-41.57	0.03
Scalar invariance	139.11	10	<0.001	59.42	0.10

Note: Estimates obtained from Mplus using MLR estimator. SBIC = Schwarz-modified Bayesian information criterion; RMSEA = root mean sqaure error of approximation.

ish respondents. This weighting should be kept in mind when interpreting the tests and caution is warranted when there are quite uneven distributions of cases across groups. For our example, we find that the chi-square test for the measurement model that combines both groups is nonsignificant and thus conclude that configural invariance holds. If we had found evidence that the model imposing configural invariance did not fit with the data, then we would need to explore different measurement model specifications across groups.

The next step in the sequence of tests involves constraining the respective factor loadings to be equal across groups. In our case, this means that latent democracy has the same effect for German and Spanish respondents on each of the indicators (i.e., $\lambda_2^{(g)} = \lambda_2^{(s)}$, $\lambda_3^{(g)} = \lambda_3^{(s)}$, and $\lambda_4^{(g)} = \lambda_4^{(s)}$). This tests what is commonly referred to as *metric invariance*.[1] The selection of the scaling indicator is an important consideration for this test (and subsequent tests). If an analyst chooses a scaling indicator that is noninvariant across groups, then it will be difficult to detect this due to fixing the factor loading to one and, as discussed below, it can be difficult to assess partial measurement invariance in the other indicators. Due to this issue, analysts should attempt to select an indicator that is most likely to be invariant across groups as their scaling indicator. In our example, we selected *free* as our scaling indicator based on this consideration.

Table 5.3 reports selected model fit statistics for the model imposing metric invariance constraints. We see that although the chi-square test statistic is significant, the other model fit statistics indicate reasonable fit with the data. The nested chi-square test, applying the scaling correction factor with the robust estimator, between the model imposing configural invariance and

---

[1] *Weak factorial invariance* and *equal factor loadings* are two other terms used to reference this constraint.

the model imposing metric invariance, however, is statistically significant. From a traditional standpoint, we would thus reject metric invariance, but based on the other model fit statistics an analyst might make a reasonable case for metric invariance depending on the context.

Our estimates with standard errors of the factor loadings constrained to be equal across groups are 0.97 (0.05), 0.93 (0.05), and −0.73 (0.05), respectively, for *civil*, *women*, and *religious*. As expected, these estimates fall between the unconstrained estimates and have slightly lower standard errors due to the combined sample size.

In most cases, the final step in the sequence of tests involves constraining the respective intercepts to be equal across groups. In our case, this means setting $\alpha_2^{(g)} = \alpha_2^{(s)}$, $\alpha_3^{(g)} = \alpha_3^{(s)}$, and $\alpha_4^{(g)} = \alpha_4^{(s)}$. This tests what is commonly referred to as *scalar invariance*.[2] Not surprisingly given our initial estimates of the intercepts from our baseline models (see Table 5.2), the model imposing scalar invariance has a poor fit with the data. In addition, the nested chi-square test, again applying the scaling correction factor, between this model and the model imposing metric invariance is statistically significant. Therefore, we conclude that our model does not exhibit scalar invariance with implications discussed below.

It is also possible to take the sequence of tests one step further and test whether constraining the error variances (and any covariances among the errors) to be equal across groups is consistent with the data. There may be some contexts where such a test is informative (e.g., if analysts are interested in exploring reliabilities of indicators across groups), but, in general, this is considered an overly restrictive form of measurement invariance and one that is not necessary to establish for most potential applications. For instance, comparing means and variances of latent variables or constructing scales remain valid even with unequal error variances across groups.

One of the main purposes of assessing measurement invariance is to analyze differences in latent means, variances, and covariances across groups. If configural and metric invariance holds, then tests for the equality of latent variances or covariances across groups are valid. If, in addition, scalar invariance holds, then tests for the equality of latent means are valid. Multiple-groups CFA offers the same approach for conducting such tests as for assessing measurement invariance: constrain the desired parameters (e.g., the latent means) to be equal across groups and test whether a model with this constraint fits the data or does not have a significantly worse fit with the data

---

[2] *Strong factorial invariance* and *equal intercepts* are two other terms used to reference this constraint.

than an unrestricted model that does not impose the equality constraint. In our example, one could make the case for metric invariance but not for scalar invariance, thus a test of the difference in the mean of latent democracy for German as compared with Spanish respondents would not be valid (i.e., any difference in the latent means will in part reflect differences in the measurement properties of the indicators).

In some cases in which metric or scalar invariance does not hold, however, it is possible to salvage analyses of differences in latent means, variances, and covariances across groups by investigating *partial measurement invariance* (Byrne et al., 1989). The tests of measurement invariance we have described so far are omnibus tests for complete sets of parameters. In contrast, an assessment of partial measurement invariance involves a more granular approach that considers constraints on individual parameters across groups. Such an approach runs the risk of overfitting the data and comes with the same issues discussed with model respecification in Chapter 4; however, if theoretically or substantively guided and employed judiciously, then it can be a useful tool for permitting tests of differences in latent means, variances, and covariances that would not otherwise be valid.

In our example, we will maintain metric invariance, and we might suspect that German and Spanish respondents' assessments of the essentialness of free elections and civil rights should be similar, while differences are more likely for assessments of the rights of women and the role of religious authorities. Therefore, we consider a model that invokes partial measurement invariance for the intercepts for *free* and *civil* along with metric invariance. Imposing these constraints results in a model that has moderate fit with the data. The chi-square test and the nested test against the model with metric invariance are both statistically significant with p values <0.001, the RMSEA is 0.06, the CFI and TLI are 0.96 and 0.94, respectively, and the SBIC is -18.59. In some research contexts, this may be sufficiently good fit to move forward with evaluating the difference in the means for latent beliefs about democracy between German and Spanish respondents. In our example, we have the mean of latent democracy for Spanish respondents constrained to 0 and estimate the mean for German respondents to be 0.51 with a standard error of 0.07. Furthermore, given our previous results for metric invariance, we can have a bit more confidence in evaluating differences in the variance for latent beliefs about democracy among German and Spanish respondents. Our estimate for the variance among Spanish respondents, 2.25, is almost twice that of the variance among German respondents, 1.23. Take together, these results suggest that Germans have higher baseline levels of beliefs about democracy but with less variance than Spaniards.

To summarize, multiple-groups CFA provides a framework for a complete assessment of measurement invariance. The most frequently recommended approach rests on the following sequence of nested tests:

1. Evaluate fit of measurement models separately for each group

2. Conduct simultaneous fit of measurement models in each group (configural invariance)

3. Conduct test of factor loadings constrained to be equal across groups (metric invariance)

4. Conduct test of intercepts constrained to be equal across groups (scalar invariance)

If metric invariance holds, then further tests of group-based differences in variances and covariances of latent variables are valid. If in addition scalar invariance holds, then further tests of group-based differences in latent means are also valid. In the event that metric or scalar invariance do not hold, then analysts may wish to explore partial measurement invariance.

5.2 MIMIC Models

The second approach to assessing measurement invariance involves an extension to CFA that permits the inclusion of variables that are not indicators of a latent variable. This approach draws on a well-known class of structural equation models, multiple-indicator multiple-outcome or MIMIC models (Jöreskog and Goldburger, 1975). MIMIC models specify one or more variables as predictors of a latent variable. These variables can be indicators of group membership or continuous variables. In order to assess measurement invariance in this framework, the standard MIMIC model is extended to permit variables that have direct effects on a latent variable to also have direct effects on one or more indicators of the latent variable (B. O. Muthén, 1985).

The MIMIC model approach to assessing measurement invariance is more commonly associated with item response theory (IRT; see Chapter 6 for a discussion) than with CFA and comes with its own jargon. In the IRT tradition, analysts are primarily concerned with whether specific indicators perform differently as measures of a latent variable across groups. In a canonical example, the latent variable is a general measure of ability (e.g., reading or math ability) and the indicators are questions from a test. The issue lies in whether a given test question is a biased measure of ability for some groups or subpopulations. For instance, at the same level of reading ability, girls and boys should have the same probability of getting the correct answer on a reading test question. To the extent that this is not the case, the reading test question is biased. This is referred to as *differential item functioning* (DIF) or *item-level bias* and can be detected by assessing the direct relationships of group membership on indicators net of any effect group membership has on latent ability.

In contrast to the types of measurement invariance outlined with multiple-groups CFA, within IRT a MIMIC model-based approach instead distinguishes between *uniform* and *nonuniform* DIF at the item level. Uniform DIF refers to the situation in which the intercept, but not the factor loading, varies across groups for a given indicator. Nonuniform DIF allows for the factor loading, as well as the intercept, of a given indicator to vary across groups. If an indicator exhibits either uniform or nonuniform DIF, then it is not invariant across groups.

There is less consensus regarding procedures for examining measurement invariance with a MIMIC model approach than with a multiple-groups CFA approach. Rather than proceed through a sequence of nested tests as in multiple-groups CFA, analysts have the option of testing for DIF in individual indicators or sets of indicators and relying on overall model fit statistics, nested tests, or tests for specific parameters to identify indicators that perform differently across groups. In keeping with our general approach of emphasizing theoretical and substantive bases for model specification and testing, we recommend a guided approach to exploring DIF that focuses on indicators with some expectation of potential differential performance across groups.

With our example involving beliefs about democracy, we might imagine that cultural differences between Germany and Spain could lead to DIF in the *women* indicator. Figure 5.2 illustrates two MIMIC models to explore uniform and nonuniform DIF for this indicator. We have reorganized the measurement model such that the indicators appear to the right rather than below the latent variable, and we have placed *women* as our last indicator to facilitate illustrating DIF. In both models, we include a binary variable that distinguishes our two groups, German and Spanish respondents, that predicts latent beliefs in democracy. For uniform DIF (illustrated in the upper panel), this binary variable also has a direct effect on the *women* indicator. For nonuniform DIF (illustrated in the lower panel), we allow the binary group variable to alter the factor loading for *women*. This is equivalent to specifying an interaction between the group variable and the latent democracy variable on the indicator for *women* and is depicted as a direct arrow from the group variable to the directed arrow from latent democracy to the indicator. This statistical interaction permits the relationship between latent

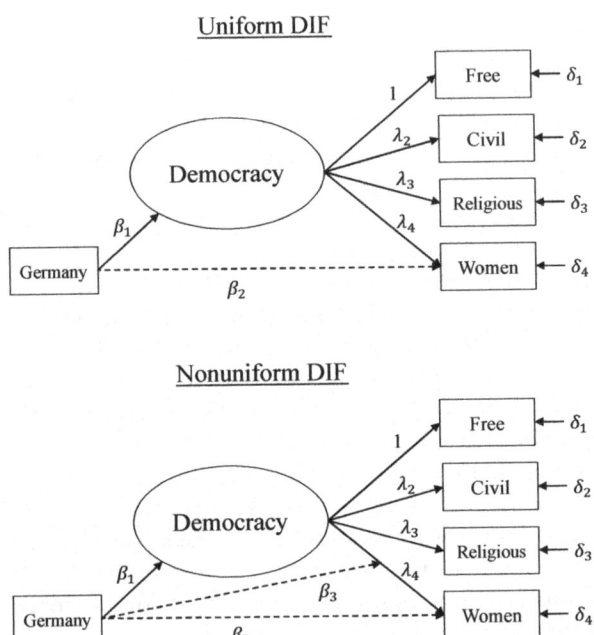

Figure 5.2: Multiple-indicator multiple-cause models for beliefs in democracy in Germany and Spain. *Note:* DIF = differential item functioning.

democracy and *women* as given by the factor loading to vary for Germany and Spain (see Fox, 2016, for a general discussion of statistical interactions). The dashed lines in the figure emphasize the effects that we are testing to determine uniform and nonuniform DIF.

Given that we are exploring DIF in a single indicator based on a substantive expectation, we examine parameter estimates for β_1 from the MIMIC model for uniform DIF and β_1 and β_2 from the MIMIC model for nonuniform DIF. If instead we decided to explore DIF in multiple indicators simultaneously, then we would rely on overall model fit statistics and tests from nested models.

Table 5.4 reports estimates and standard errors for selected parameters from a baseline MIMIC model and the MIMIC models permitting uniform and nonuniform DIF in the *women* indicator. In the uniform DIF MIMIC model, we find limited evidence of a difference in intercepts in the indicator across German and Spanish respondents with an estimate of -0.12

Table 5.4: Selected parameter estimates from the democracy multiple-indicator multiple-cause models.

	Baseline	Uniform DIF	Nonuniform DIF
women intercept	8.78 (0.05)	8.83 (0.06)	8.83 (0.06)
women factor loading	0.87 (0.04)	0.90 (0.05)	0.98 (0.05)
β_1	0.66 (0.07)	0.70 (0.07)	0.70 (0.07)
β_2	—	-0.12 (0.07)	-0.05 (0.09)
β_3	—	—	-0.19 (0.08)

Note: Unstandardized estimates with standard errors in parentheses. Parameters refer to models illustrated in Figure 5.2. Estimates were obtained from Mplus using MLR estimator. DIF = differential item functioning.

and a standard error of 0.07. In the nonuniform DIF MIMIC model, however, we do find some evidence of a difference in factor loadings across groups with an estimate for the interaction term of -0.19 and a standard error of 0.08. This suggests that there is a weaker relationship between latent beliefs in democracy and the *women* indicator among German than among Spanish respondents. Notice that this is consistent with the estimates of the factor loadings from separate measurement models for Germany and Spain reported in Table 5.2.

Based on this analysis of measurement invariance (or DIF) for the *women* indicator, we can conclude that there is no evidence of uniform DIF but some evidence of nonuniform DIF. Depending on the research context, this could be a substantive finding or a source of concern for subsequent use of the indicators as components of a scale. In addition, if we are interested in comparing latent means across groups, then we can interpret the estimate for β_1 having adjusted for nonuniform DIF in the *women* indicator. In this case, our estimate is 0.70, which indicates a higher mean level of beliefs about democracy among German respondents as compared with Spanish respondents. Note that this estimate is similar to what we found allowing for partial measurement invariance in the multiple-groups CFA context and rests on the same assumptions (i.e., that measurement invariance holds for at least one of the nonscaling indicators).

In summary, the use of modified MIMIC models to assess measurement invariance provides an alternative to the multiple-groups CFA approach that holds some advantages. The MIMIC model approach permits a broader exploration of potential sources of measurement invariance through the introduction of group variables with more categories than is typically feasible in a multiple-groups CFA context or even continuous variables. In addition, the MIMIC model approach tends to be more focused and theoretically or substantively driven to identify uniform or nonuniform DIF in individual indicators. The MIMIC model approach, however, does not provide a holistic assessment of measurement invariance and enjoys less of a consensus in how to organize an exploration of measurement invariance than the standard sequence of tests in the multiple-groups CFA approach.

5.3 Conclusion

In this chapter, we outlined two approaches to assessing the performance of measurement models and measurement invariance across groups. The two approaches have offsetting strengths and weaknesses. The multiple-groups approach permits a holistic assessment of measurement invariance for a typically small number of discretely defined groups. The MIMIC model approach permits an analysis of measurement invariance for a larger number of groups or subpopulations defined by continuous variables, but the assessment is typically focused on indicator by indicator uniform and nonuniform DIF.

For many analyses, the goal of evaluating measurement invariance is to find that measurement models have largely the same performance across groups. If so, this permits an analysis of group-based differences in scale or latent variable means, variances, and covariances that is not confounded by artifacts of measurement. Such an interpretation, however, rests on the assumption that the set of indicators (or instrument or scale) as a whole function the same way across groups. In the IRT tradition, this idea is referred to as differential test functioning (DTF). In most contexts, it is not possible to test for DTF, but it should be given some consideration when researchers are examining latent variables (or scales) across groups.

5.4 Further Reading

For readers interested in learning more about differential item function, a fuller treatment can be found in Osterlind and Everson (2009). A consideration of the issue of measurement invariance from the IRT tradition (but which contains much of value for a CFA analyst) may be found in Wilson (2005). For another treatment of the use of MIMIC models to assess measurement invariance, see Woods (2009) and Woods and Grimm (2011). For an exemplar assessing measurement invariance in a cross-national context, see Cieciuch et al. (2018).

CHAPTER 6. CATEGORICAL INDICATORS

Our discussion so far has focused on CFA models in which all the indicators are assumed to be continuous (albeit potentially nonnormal) variables. In many cases, however, analysts encounter categorical variables of various types (e.g., dichotomous or binary, ordered categorical or ordinal, and unordered categorical or nominal). For instance, a set of items measuring gender attitudes may include responses ranging from 1 "strongly disagree" to 4 "strongly agree," or a series of binary variables indicating the presence or absence of symptoms may be used as measures of an underlying disorder. As noted in Chapter 3, treating such variables as continuous can be an option, particularly with robust ML estimators. Furthermore, simulation studies find that treating ordinal measures with five or more categories and roughly symmetrical distributions as continuous often introduces minimal bias in model fit statistics or parameter estimates (Rhemtulla et al., 2012). Nonetheless, an understanding of CFA with categorical indicators is valuable when researchers are working with ordinal measures with fewer than five categories, with skewed or otherwise nonnormal distributions, or when researchers have an interest in specifically addressing the discrete nature of the measures.

In this chapter, we provide an overview of CFA with categorical indicators. In particular, we focus on ordinal indicators with an understanding that dichotomous indicators are a special case with only two values. We begin with a discussion of two ways to conceptualize ordinal indicators and specify measurement models. We then present two commonly used estimators for CFA models with categorical indicators and discuss options for assessing model fit and interpreting the parameter estimates. We conclude with a brief discussion of the relationship between CFA with categorical indicators and IRT, a closely related class of models.

Throughout this chapter, we work with a measurement model for Christian nationalism in the United States developed using data from the 2017 wave of the Values and Beliefs of the American Public Survey (Whitehead et al., 2018). These data include six potential measures of Christian nationalism for 1,378 respondents (see Table 6.1). For the examples in this chapter, we recode the measures to combine "agree" and "strongly agree" as well as "disagree" and "strongly disagree," and we also treat undecided as a middle category. With these recodes, each measure takes three values: 1 = *strongly disagree or disagree*, 2 = *undecided*, and 3 = *agree or strongly agree*.

Table 6.1: Item prompts and responses for Christian nationalism indicators.

	SD or D	U	A or S
Please rate the extent to which you agree or disagree with the following statements:			
x_1: The federal government should declare the United States a Christian nation (*nation*).	1	2	3
x_2: The federal government should advocate Christian values (*values*).	1	2	3
x_3: The federal government should enforce strict separation of church and state (*separate*).	1	2	3
x_4: The federal government should allow the display of religious symbols in public spaces (*symbols*).	1	2	3
x_5: The success of the United States is part of God's plan (*plan*).	1	2	3
x_6: The federal government should allow prayer in public schools (*prayer*).	1	2	3

Note: SA = strongly agree, A = agree, D = disagree, SD = strongly disagree, U = undecided.

Figure 6.1 illustrates the distributions for each of the indicators. We see that none of the indicators approximate a symmetric distribution. Across all of the indicators, the proportion of respondents reporting undecided is lower than for either disagree or agree. Given the small number of response categories and the bimodal distributions of the indicators, a CFA model addressing the categorical nature of the measures is called for.

6.1 Conceptualization of Ordinal Measures

There are two broad conceptual approaches to working with ordinal measures. The first approach involves treating the ordinal measure as a crude indicator of an underlying continuous variable. For instance, as depicted in Figure 6.2, we might imagine that there is a continuum of how much support people feel for the statement that the federal government should advocate Christian values. In Figure 6.2, this continuum ranges from low to high and is depicted as having a normal distribution in the population. Our survey item, however, does not capture the full continuum but rather three responses (disagree, undecided, and agree) that correspond to regions of the continuum demarcated by two thresholds. For example, respondents who lie below the first threshold on the continuum report that they disagree with the statement.

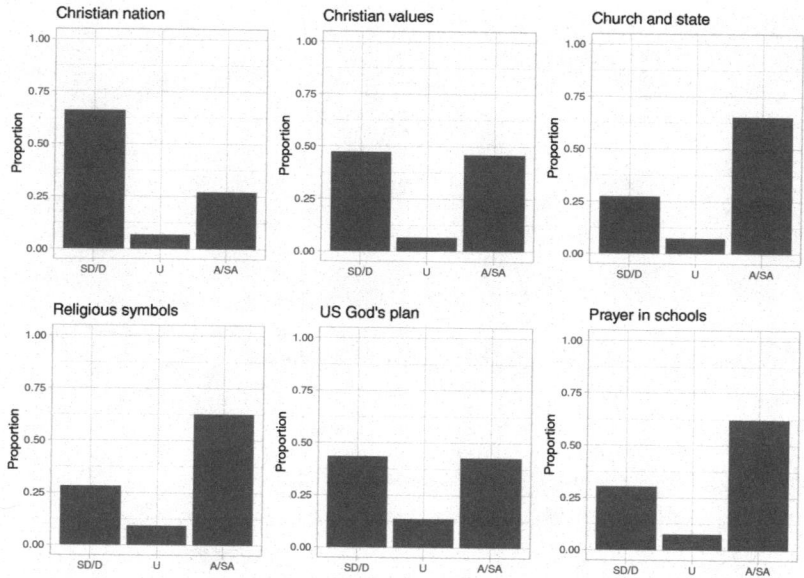

Figure 6.1: Distributions of indicators of Christian nationalism.
Note: SD/D = strongly disagree/disgaree, U = undecided, A/SA = agree/strongly agree.

A key assumption underlying this approach is that the values of the observed indicator are in fact ordered and lie along a single dimension. For our indicators of Christian nationalism, we might question this assumption on two grounds. First, we have coded "undecided" as a middle category that lies between "disagree" and "agree." This is a common approach, but it is reasonable to consider whether some people might respond "undecided" for reasons that would not place them between "disagree" and "agree" (e.g., perhaps they simply did not understand the statement). Second, it is possible that the responses, and in particular the original range with five categories, reflect two dimensions rather than one. One dimension might capture a continuum of support for the statement, as noted above, while another dimension might capture intensity of beliefs (e.g., a continuum ranging from tepid to strong feelings in either direction). If either the undecided category is not ordered or the responses reflect more than one dimension, then a more complex measurement model that does not assume ordinality for the indicators is needed for Christian nationalism.

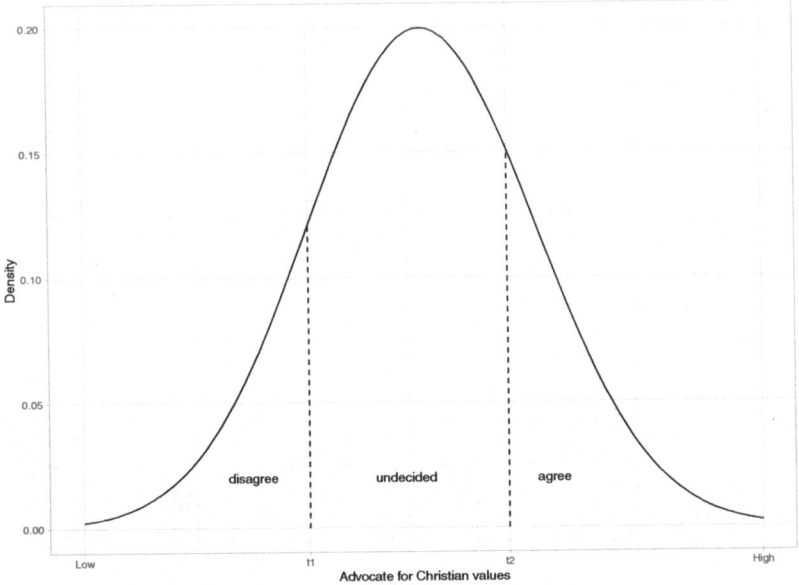

Figure 6.2: Illustration of underlying continuum for ordinal measure.

Rather than invoking the idea of an underlying continuous variable, the second approach instead models the probabilities of the responses as given. For instance, rather than viewing the support people feel for the idea the federal government should advocate Christian values as lying on a continuum, we instead model the probabilities of agreeing, being undecided, or disagreeing with the sentiment. The same assumption concerning the ordering of values along a single dimension applies with this approach as well.

Figure 6.3 illustrates the difference between the two conceptual approaches when applied to a CFA measurement model. In both panels, the dotted arrows signify nonlinear relationships due to the categorical nature of the observed indicators and the dashed lines denote the presence of the other four indicators. Panel A depicts a model in which the latent variable for Christian nationalism is measured by the underlying continuous variables for each of the observed indicators. The underlying continuous variables are then connected to the observed indicators via a nonlinear function as indicated by the dotted arrows. Panel B, in contrast, illustrates a model in which the latent variable for Christian nationalism has direct nonlinear relationships with the observed indicators.

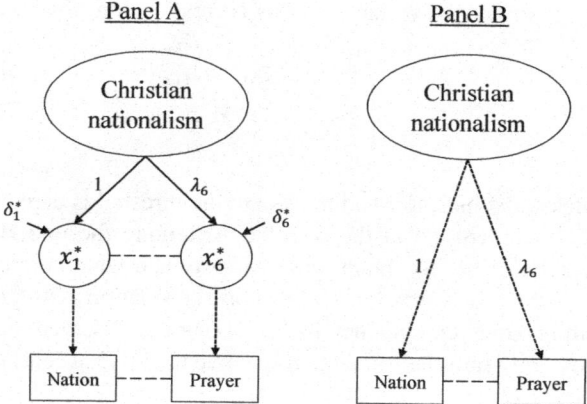

Figure 6.3: Illustration of confirmatory factor analysis for Christian nationalism.

Both approaches to conceptualizing ordinal measures lead to identical model fit and parameter estimates. The approach involving underlying continuous variables for the observed indicators is more commonly invoked in the CFA tradition while directly modeling the probabilities of responses for the observed indicators is more commonly invoked in the IRT tradition.

6.2   Estimators

Analysts have two broad choices of estimators with different strengths and weaknesses when fitting measurement models with ordinal indicators. Currently, the most common choice in the CFA tradition is a variant of a weighted least squares (WLS) estimator. The main alternative is an ML estimator derived specifically for ordinal (or dichotomous) indicators that is standard in IRT modeling. In this section, we provide a brief overview of each before illustrating their use with a measurement model for Christian nationalism.

6.2.1   Polychoric Correlations

Before turning to the WLS estimator, it is important to understand how we can capture the relationships between ordinal measures. From Table 6.1, we see that x_1 to x_6 represent our six observed ordinal measures of Christian nationalism. Suppose that x_1^* to x_6^* represent the presumed underlying continuous variables for each of the respective measures. The relationship between an underlying continuous variable and its corresponding observed variable is given by

$$x_j = \begin{cases} 1 & -\infty \leq x_j^* < \tau_{1j} \\ 2 & \tau_{1j} \leq x_j^* < \tau_{2j} \\ 3 & \tau_{2j} \leq x_j^* < \infty, \end{cases} \qquad (6.1)$$

where j indexes the indicator and the τs are the thresholds depicted in Figure 6.1. We are interested in the correlations among the underlying continuous variables (x^*s). To obtain these, we estimate what is referred to as polychoric correlation matrix.[1] The polychoric correlation matrix uses information from bivariate contingency tables among the observed indicators to estimate the correlations between the respective underlying continuous variables (Olsson, 1979).

Table 6.2 reports the Pearson correlations among the observed measures of Christian nationalism below the diagonal and the polychoric correlations above the diagonal. We see that in this case, and as is true in general, treating the indicators as continuous and estimating Pearson correlations substantially underestimates the relationships among the underlying continuous variables as given by the polychoric correlations. For instance, the Pearson correlation between the first two indicators, *nation* and *values*, is 0.62 while the polychoric correlation for the underlying continuous variables of the first two indicators is 0.86.

6.2.2 Weighted Least Squares Estimator

In the CFA tradition, the most commonly used estimator for models with ordinal (or dichotomous) indicators is a variant of a WLS estimator.[2] As with ML estimators for continuous indicators, WLS estimators seek to minimize the difference between the model-implied covariance matrix, $\Sigma(\theta)$, and the observed covariance matrix, S, except that the polychoric correlation matrix stands in for the observed covariance matrix. In addition, WLS

---

[1] There are a number of different terms to reference specific combinations of variables. A correlation between two binary variables is referred to as a tetrachoric correlation; between a continuous variable and a binary variable is referred to as a biserial correlation; and between a continuous variable and an ordinal variable is referred to as a polyserial correlation. For ease of presentation, we use the term *polychoric* to reference all possible combinations of correlations involving an ordinal variable.

[2] An earlier approach to fitting CFAs with categorical indicators involved substituting the polychoric correlation matrix for the observed correlation matrix in the standard ML estimator. This approach leads to a consistent estimator for the parameters but incorrect standard errors and model fit statistics (Bollen, 1989).

Table 6.2: Pearson and polychoric correlations among Christian nationalism indicators.

	x_1	x_2	x_3	x_4	x_5	x_6
x_1		0.86	−0.33	0.58	0.70	0.71
x_2	0.62		−0.39	0.65	0.69	0.72
x_3	−0.20	−0.25		−0.28	−0.40	−0.51
x_4	0.36	0.45	−0.17		0.56	0.73
x_5	0.49	0.52	−0.27	0.39		0.70
x_6	0.43	0.51	−0.31	0.55	0.51	

Note: Pearson correlations among the observed indicators (x_1 to x_6) are reported below the diagonal and polychoric correlations among the underlying continuous variables (x_1^* to x_6^*) are reported above the diagonal. Correlations were obtained from Stata.

estimators adjust for the nonnormality of the indicators by incorporating a weight matrix in the fit function that is a function of the asymptotic covariance matrix of the polychoric correlations (i.e., the variances and covariances of the estimates of the polychoric correlations).

The asymptotic covariance matrix of the polychoric correlations, however, can be quite large and lead to problems with estimation. For the six indicators of Christian nationalism, there are 21 nonredundant elements of the polychoric correlation matrix, and the asymptotic covariance matrix is thus 21×21 and has 231 nonredundant elements. The size of the asymptotic covariance matrix substantially increases with the number of indicators. In a measurement model involving 20 indicators, such as a typical model for personality traits, the polychoric correlation matrix has 210 nonredundant elements that leads to an asymptotic covariance matrix with 22,155 nonredundant elements to estimate.

The WLS estimator requires large sample sizes to obtain accurate estimates of the elements of the asymptotic covariance matrix, and the estimator involves inverting the asymptotic covariance matrix, which can also pose numerical problems. In part due to these issues, the standard WLS estimator has not performed well in simulation studies when used with ordinal indicators (Flora and Curran, 2004).

To help mitigate these issues, psychometricians developed variants of the WLS estimator that ease the computational burden. One of the most popular current variants is the WLS-MV estimator that substitutes the diagonal of the asymptotic covariance matrix for the full matrix as the weight matrix in obtaining parameter estimates and reports robust standard errors along with a mean- and variance-adjusted chi-square test statistic (Muthén and Muthén, 2017).[3] The use of the diagonal of the asymptotic covariance matrix substantially reduces the number of nonredundant elements to estimate and avoids numerical issues that come with inverting the full asymptotic covariance matrix. These features allow analysts to fit CFA measurement models with categorical indicators with relatively small samples. A sample size of roughly 200 may be sufficient for models involving up to 10 to 15 indicators, although more simulation studies are needed to provide guidance in this area.

An identification issue arises when using the polychoric correlation matrix with the WLS-MV estimator. As discussed above, the polychoric correlation matrix captures the relationships between the underlying continuous variables (x^*s), and we need to assign a scale to these variables in the same way that we need to scale any latent variable (see discussion in Chapter 3). There are two conventional options for doing so in the CFA framework. The first, and most frequently used, option involves fixing the variance of each x^* to 1. An implication of this identification constraint is that error variances for the categorical indicators are not free parameters to be estimated. In Mplus, this option is referred to as the *delta parameterization* and is the default option. The second option involves fixing the error variances for the categorical indicators to 1, in which case the variances of the x^*s are not free parameters to be estimated. In Mplus, this option is referred to as the *theta parameterization*, and it is more closely aligned with IRT model specification. The two options lead to equivalent measurement models and have no impact on testing overall model fit or nested tests, thus the choice between the two depends on whether or not the analyst is interested in working with estimates of error variances.

As noted above, the WLS-MV estimator produces a mean- and variance-adjusted chi-square test statistic that can be used to assess overall model fit in the same manner discussed for models with continuous indicators. In addition, a range of model fit indices (e.g., the CFI and TLI) and the RMSEA are available. Information criterion (AIC, BIC, SBIC, or variants) that are based on log-likelihoods or deviances, however, are not available for WLS-MV estimators. One additional wrinkle with WLS-MV estimators, as with other robust estimators, is that the difference in chi-squares between two

---

[3]The WLS-MV estimator is available in Mplus software (Muthén and Muthén, 2017). Similar variants, such as the diagonally weighted least squares (DWLS) estimator, are available in other statistical software packages (e.g., both LISREL and the lavaan package in R).

nested models is not itself chi-square distributed (Satorra, 2000). To conduct chi-square tests of nested models, analysts must apply a correction factor, a process that is automated in Mplus (Muthén and Muthén, 2017).

The standard parameter estimates that we obtain from a WLS-MV estimator include the factor loadings that capture the relationship between the latent variables and the underlying continuous variables (x^*s), the thresholds that link the underlying continuous variables to the observed indicators, and the variances and covariances among the latent variables. By convention, the means of the latent variables and the intercepts for the underlying continuous variables are fixed to 0 for model identification. The WLS-MV estimator relies on a probit link, so the factor loadings capture the relationship between the latent variables and the underlying continuous variables in the probit metric.

In some contexts, we might be interested in the relationships between the latent variables and the observed indicators themselves as opposed to their underlying continuous variables. With the estimates of the factor loadings and the thresholds, we can calculate the effects of latent variables on the probabilities of the response categories for each of the indicators. As we will demonstrate below, these are relatively straightforward calculations and can provide a more intuitive basis for interpretation than the probit metric.

6.2.3 Maximum Likelihood Estimator

In the CFA context, WLS estimators are sometimes referred to as *limited-information estimators* in that they rely on a summary of the data, in particular, the polychoric correlation matrix for estimates.[4] In contrast, ML estimators for categorical indicators are sometimes referred to as *full-information estimators* as they work with the raw data rather than summary statistics such as a polychoric correlation matrix.[5] As such, ML estimators for measurement models with categorical indicators have the advantage of theoretically greater efficiency (i.e., more precise estimates) than the various WLS estimators, though they still typically require larger sample sizes than comparable measurement models with continuous indicators.

ML estimators obtain parameter estimates by maximizing the likelihood of the responses patterns for the indicators across all the cases. For an esti-

---

[4] The terms *limited-information* and *full-information* are used in other areas of statistics to mean different things (e.g., to distinguish between estimators for a subset of model equations versus a complete set of model equations).

[5] Note that this is a different form of an ML estimator than the one discussed in Chapter 3 for continuous indicators.

mator derived from the logit link, the probability of observing a response c or greater for indicator j of latent variable ξ is given by

$$\Pr[x_{ij} \geq c] = \frac{1}{1 + \exp(\tau_j - \lambda_j \xi)}, \tag{6.2}$$

where τ is a threshold as described above and λ is a factor loading. Similarly, for an estimator derived from the probit link, the probability of observing a response c or greater for indicator j of latent variable ξ is given by

$$\Pr[x_{ij} \geq c] = \Phi(-\tau_j + \lambda_j \xi), \tag{6.3}$$

where $\Phi(.)$ is the standard normal cumulative distribution function. Note the absence of an error term in both equations. This is a standard identification constraint that is an analog of the identification constraints discussed with the WLS-MV estimator. The link function has little impact on the assessment of model fit and the parameter estimates can be translated from the logit to the probit metric and vice versa. In practice, the choice of which link function to use largely depends on the conventions in the particular area of applied work or the default setting in the particular statistical software package used for fitting CFAs.

Although the ML estimator for categorical indicators is theoretically more efficient than the WLS-MV estimator and does not require the intermediate step of estimating a polychoric correlation matrix for the indicators, it does have some disadvantages. First, maximizing the likelihood function relies on numerical integration that can be computationally intensive, particularly for models involving multiple latent variables. Second, ML estimators for categorical indicators produce a different form of a chi-square test statistic for overall model fit than WLS-MV estimators. In this context, the chi-square test statistic with ML estimators is based on observed versus expected frequencies in the multiway frequency tables among the observed indicators (Agresti, 2002). This form of a chi-square test performs poorly and should not be interpreted when there are cells with zero or near-zero counts, a common occurrence in this context. In addition, this form does not lend itself to the various fit indices that have been developed based on the chi-square test statistic for models with continuous indicators. As such, there are not as many options for assessing the overall model fit and testing the nested models with the ML estimator as compared with the WLS-MV estimator.

6.2.4 Parallel Lines Assumption

The parallel lines assumption, also known as the proportional odds assumption with the logit link, referenced for ordinal regression models holds for both the WLS-MV and the ML estimators (Long, 1997).[6] In the CFA context, the parallel regression assumption concerns the factor loadings. In CFA models with ordinal indicators, the effect of a latent variable on an indicator is constrained to be equal across thresholds (i.e., only a single factor loading is estimated for each indicator as opposed to one for each threshold). For example, we assume that the effect of Christian nationalism on support for the statement that the federal government should advocate Christian values is the same in moving people from "disagree" to "undecided" as in moving people from "undecided" to "agree." With ML estimators, this assumption can be relaxed and tested (see Long (1997 for options in a regression context), although it is not often done.

6.3 Model Fit and Parameter Interpretation

Figure 6.4 illustrates our baseline measurement model for Christian nationalism. In this model, all six measures are treated as ordinal indicators of a single latent variable for Christian nationalism. In the model, we explicitly illustrate the relationship between the latent variable and the underlying observed continuous variables (the x^*s) as well as the nonlinear relationship between the underlying continuous variables and the observed ordinal indicators. In most published work, however, the measurement models are depicted without the underlying continuous variables, and the nonlinear relationships are not explicitly illustrated. This information is more typically provided in the data and methods section.

6.3.1 Model Fit

We begin with an assessment of model fit for our measurement model for Christian nationalism. Table 6.3 provides a comparison of model fit statistics from several different estimators. As noted above, the choice of delta or theta parameterization with the WLS-MV estimator has no effect on the model fit statistics. For the WLS-MV estimators, we see a statistically significant chi-square test statistic that is indicative of poor model fit. The CFI and TLI around 0.98 indicate reasonable model fit, while the RMSEA of 0.09 is a bit high (see Chapter 4 for a discussion). We see quite similar estimate for the ML estimator across the probit and logit links. In both cases, the chi-square tests are significant indicating poor fit, which is consistent with the

---

[6]There are forms of ordered regression models that relax the parallel lines assumption (Fullerton, 2009), but these are not commonly used in the CFA context.

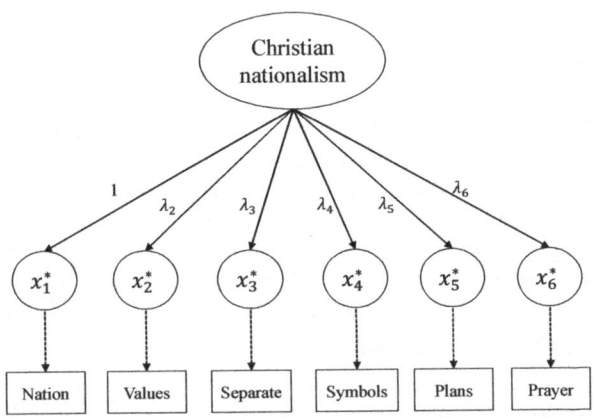

Figure 6.4: Measurement model for Christian nationalism.

WLS-MV chi-square test, even though it is a different form of the chi-square test. We note, however, that some cells in the multiway contingency table were zero or near-zero, and these chi-square tests are likely inaccurate. As a point of comparison, in the final column we include the fit statistics from treating the indicators as continuous and using a robust maximum likelihood estimator as outlined in Chapter 3. With the MLR, we see a similar pattern in the measures of model fit as with the WLS-MV estimator, but a bit worse in each case.

For our measurement model for Christian nationalism, the fit statistics for all estimators point to concerns with model misspecification. At this stage, we would consider substantively based alternative model specifications as outlined in Chapter 4. For instance, we might consider checking for poor performing indicators and removing them, considering whether there might be more than one latent variable, or treating the indicators as nominal rather than ordinal, among other theoretically or substantively informed possibilities. For the purposes of our illustration, however, we continue with interpreting estimates from our baseline measurement model.

6.3.2 Parameter Estimates

We begin with parameter estimates from WLS-MV estimators. Table 6.4 reports selected parameter estimates from both the WLS-MV estimator using the delta parameterization that fixes the variances of the x^*s to 1 and the theta parameterization that fixes the variances of the residuals to 1. The first panel includes unstandardized and standardized factor loadings. The

Table 6.3: Model fit statistics for Christian nationalism measurement model.

	WLS-MV		ML		MLR
	Delta	Theta	Probit	Logit	
Chi-square	111.58	111.58	3498.44	3569.10	159.86
df	9	9	705	705	9
p Value	<0.001	<0.001	<0.001	<0.001	<0.001
CFI	0.99	0.99			0.93
TLI	0.98	0.98			0.88
RMSEA	0.09	0.09			0.11

Note: WLS-MV Delta and Theta refer to the two different parameterizations discussed in the section on WLS-MV estimators. ML Probit and ML Logit refer to ML estimators for categorical indicators using the probit and logit links, respectively. MLR refers to a robust maximum likelihood estimator treating the indicators as continuous. The chi-square test statistic differs in form across the estimators as discussed in the section on ML estimators. Model fit statistics were obtained from Mplus. WLS-MV = weighted least squares estimator–means and variance adjusted; ML = maximum likelihood; df = degrees of freedom; CFI = comparative fit index; TLI = Tucker–Lewis index; RMSEA = root mean square error of approximation.

unstandardized factor loadings are interpreted as the effect of a 1-unit change in the latent variable on the given indicator in the probit metric. A 1-unit increase in Christian nationalism is associated with, for instance, a 0.83-unit increase in the probit of symbols in the delta parameterization or a 0.55-unit increase in the (differently scaled) probit of symbols in the theta parameterization. Such statements provide limited intuition for readers, so analysts typically focus on the direction and size of the relationship relative to the standard error if reporting unstandardized estimates (e.g., increasing Christian nationalism is associated with an increased probability of agreeing with the statement about symbols). The standardized factor loadings shift the interpretation to a standard deviation change in Christian nationalism and standardize the x^*s in the theta parameterization (they are already standardized in the delta parameterization). As we can see, the standardized loadings are identical across the two parameterizations.

In the second panel of Table 6.4, we report the estimates of the thresholds for each indicator. Given that the indicators take three values (disagree, undecided, and agree), we have two thresholds for each indicator. Much

Table 6.4: Selected parameter estimates for Christian nationalism measurement model from WLS-MV estimators.

	WLS-MV Delta		WLS-MV Theta	
	Ustd	Std	Ustd	Std
Factor Loadings				
nation	1 (—)	0.90	1 (—)	0.90
values	1.01 (0.02)	0.90	1.03 (0.13)	0.90
separate	−0.52 (0.04)	−0.46	−0.26 (0.03)	−0.46
symbols	0.83 (0.03)	0.75	0.55 (0.06)	0.75
plan	0.87 (0.02)	0.78	0.62 (0.06)	0.78
prayer	0.97 (0.02)	0.87	0.87 (0.10)	0.87
Thresholds				
nation τ_1	0.42 (0.04)		0.94 (0.10)	
nation τ_2	0.61 (0.04)		1.38 (0.12)	
values τ_1	−0.07 (0.03)		−0.15 (0.08)	
values τ_2	0.10 (0.03)		0.24 (0.08)	
separate τ_1	−0.60 (0.04)		−0.68 (0.04)	
separate τ_2	−0.39 (0.04)		−0.44 (0.04)	
symbols τ_1	−0.58 (0.04)		−0.87 (0.06)	
symbols τ_2	−0.32 (0.03)		−0.48 (0.05)	
plan τ_1	−0.16 (0.03)		−0.26 (0.05)	
plan τ_2	0.19 (0.03)		0.30 (0.05)	
prayer τ_1	−0.51 (0.04)		−1.04 (0.09)	
prayer τ_2	−0.31 (0.03)		−0.62 (0.08)	
Variance				
Christian nationalism	0.81 (0.03)	1 (—)	4.13 (0.69)	1 (—)
R^2				
nation		0.81		0.81
values		0.81		0.81
separate		0.22		0.22
symbols		0.56		0.56
plan		0.61		0.61
prayer		0.76		0.76

Note: Standard errors are in parentheses. Estimates were obtained from Mplus. WLS-MV = weighted least squares estimator–means and variance adjusted; Ustd = unstandardized estimates; Std = standardized estimates.

like the intercepts in a CFA with continuous indicators, these are rarely discussed. As we will show in the next section, however, the threshold estimates are necessary for translating the effects of the latent variable into the probability metric, which can be a useful aid to interpretation.

The third panel of Table 6.4 contains the estimated variance of the latent variable across the two parameterizations, and the fourth panel reports the R^2s. In the context of categorical indicators, these R^2s represent the amount of variance in the x^*s explained by the latent variable as opposed to the variance in the observed indicators. As with continuous indicators, this information can be useful for diagnosing potentially weak indicators. In this case, for instance, we see that latent Christian nationalism only accounts for 22% of the variance in the underlying continuous variable linked to the statement about the separation of church and state.

Table 6.5 presents the parameter estimates from the ML estimators using a probit and a logit link. We see that the ML probit estimates are quite similar to the WLS-MV estimates with the theta parameterization. The small differences can be attributed to a combination of using summary statistics versus the raw data and different numerical algorithms. The estimates from the ML probit have the same interpretation as the WLS-MV estimates.

The ML logit estimates for the factor loadings and the thresholds differ from the ML probit estimates due to the use of a different link function. The unstandardized estimates represent the effect of Christian nationalism on each of the indicators in the log-odds metric. These estimates can be exponentiated and interpreted as odds ratios (Long, 1997). For instance, exponentiating the factor loading for *values* results in an odds ratio of $\exp(0.94) = 2.56$. This indicates a 1-unit increase in latent Christian nationalism is associated with a 2.56 higher odds of increasing agreement with the statement that the federal government should advocate Christian values. In contrast, exponentiating the factor loading for *separate* results in an odds ratio of $\exp(-0.25) = 0.78$. This indicates a 1-unit increase in latent Christian nationalism is associated with a 0.22 lower odds of increasing agreement with the federal government should enforce strict separation of church and state. More generally, positive factor loadings in the log-odds metric translate into increasing likelihoods of higher values on the observed ordinal indicators and vice versa for negative factor loadings. Standardized loadings are interpreted similarly except that the scale of the latent variable is standardized such that a 1-unit change reflects a 1 standard deviation change in the latent variable.

As with the other estimators, in the CFA context the thresholds are not often interpreted directly but are instead used in calculating probabilities of specific responses for the observed indicators. We illustrate this below.

Table 6.5: Selected parameter estimates for Christian nationalism measurement model from ML estimators.

	ML Probit		ML Logit	
	Ustd	Std	Ustd	Std
Factor Loadings				
nation	1 (—)	0.90	1 (—)	0.90
values	0.98 (0.10)	0.90	0.94 (0.10)	0.89
separate	−0.25 (0.03)	−0.45	−0.23 (0.03)	−0.43
symbols	0.51 (0.06)	0.73	0.50 (0.06)	0.71
plan	0.61 (0.07)	0.79	0.59 (0.07)	0.77
prayer	0.79 (0.10)	0.86	0.79 (0.10)	0.85
Thresholds				
nation τ_1	0.96 (0.10)		1.69 (0.18)	
nation τ_2	1.40 (0.11)		2.48 (0.21)	
values τ_1	−0.12 (0.08)		−0.19 (0.13)	
values τ_2	0.26 (0.08)		0.46 (0.14)	
separate τ_1	−0.68 (0.04)		−1.12 (0.07)	
separate τ_2	−0.45 (0.04)		−0.73 (0.07)	
symbols τ_1	−0.83 (0.06)		−1.42 (0.10)	
symbols τ_2	−0.45 (0.05)		−0.76 (0.09)	
plan τ_1	−0.24 (0.06)		−0.42 (0.09)	
plan τ_2	0.32 (0.06)		0.54 (0.10)	
prayer τ_1	−0.96 (0.08)		−1.71 (0.15)	
prayer τ_2	−0.56 (0.07)		−1.00 (0.13)	
Variance				
Christian nationalism	4.33 (0.76)	1 (—)	13.83 (2.51)	1 (—)
R^2				
nation		0.81		0.81
values		0.81		0.79
separate		0.21		0.18
symbols		0.53		0.51
plan		0.62		0.59
prayer		0.73		0.72

Note: Standard errors are in parentheses. Estimates were obtained from Mplus. WLS-MV = weighted least squares estimator–means and variance adjusted; Ustd = unstandardized estimates; Std = standardized estimates.

Finally, we note that the R^2s from the ML estimator with the logit link are quite similar to the R^2s from the ML estimator with the probit link and also the WLS-MV estimator. As before, the estimates are interpreted with respect to the amount of variance explained in the underlying continuous variables for each observed indicator. In this case, *separate* continues to have a notably low R^2 of 0.18, which indicates that latent Christian nationalism explains relatively little of the variance in the underlying continuous variable for *separate*.

6.3.3 Postestimation

For most analysts, interpreting factor loadings in the probit or logit (log-odds) metric provides limited intuition about the precise relationships between latent variables and observed indicators. Transforming logit estimates into odds ratios via exponentiation aids interpretation, but more can be learned from the estimates by a further transformation into the probability metric when analysts are interested in the effects of latent variables on response categories. Interpretations in the probability metric provide analysts with an understanding of how latent variables are associated with the probabilities of responses for the observed indicators. In this section, we illustrate how to use the estimates of factor loadings and thresholds to calculate predicted probabilities of responses for observed indicators and how to calculate effects of latent variables on the observed indicators in the probability metric as opposed to the probit or logit metrics.

For an ordinal indicator x of a latent variable ξ, the predicted probability of response c is given by

$$\Pr(x = c) = \Pr(x \geq c) - \Pr(x \geq c + 1) \tag{6.4}$$

with the equations for $\Pr(x \geq c)$ and $\Pr(x \geq c + 1)$ depending on the link function as given in equations (6.2) and (6.3) for the logit and probit links, respectively, above. To illustrate these calculations, consider our estimates for *nation*, our first indicator of Christian nationalism.

In our illustration, we use the WLS-MV unstandardized estimates with the delta parameterization (see Table 6.4). The same calculations can be done using the theta parameterization estimates or either of the ML estimates and a similar pattern will emerge, though scaled differently. Plugging the WLS-MV delta parameterization estimates into the equation for the probit link, we can calculate the predicted probabilities of disagreeing (D), being undecided

(U), and agreeing (A) at the mean of 0 for Christian nationalism as

$$\Pr(nation \geq D) = 1 \tag{6.5}$$

$$\Pr(nation \geq U) = \Phi((-1)(0.42) + 1(0)) = 0.34 \tag{6.6}$$

$$\Pr(nation \geq A) = \Phi((-1)(0.61) + 1(0)) = 0.27 \tag{6.7}$$

and then

$$\Pr(nation = D) = \Pr(nation \geq D) - \Pr(nation \geq U) = 0.66 \tag{6.8}$$

$$\Pr(nation = U) = \Pr(nation \geq U) - \Pr(nation \geq A) = 0.07 \tag{6.9}$$

$$\Pr(nation = A) = \Pr(nation \geq A) - 0 = 0.27. \tag{6.10}$$

We see that at the mean level of Christian nationalism, the response with the highest predicted probability is disagree with 0.66.

We can do similar calculations for the other indicators and also choose a range of values for the latent variable. Even better, we can plot how the predicted probabilities for each response change over a range of latent Christian nationalism. Figure 6.5 illustrates the predicted probabilities of responses for all the indicators of Christian nationalism. The curves are labeled "SD/D" for disagree/strongly disagree, "U" for undecided, and "A/SA" for agree/strongly agree.

The graphs illustrate a number of interesting patterns. First, we note that, as expected given the different valence of the question, the predicted probabilities have the opposite pattern for responses to *separate* as compared with the other indicators. Second, we see that the predicted probabilities of agreeing/strongly agreeing for the *symbols* and *prayer* indicators are higher at lower levels of Christian nationalism than across the other indicators. Third, we see that the *plan* indicator has the highest peak in the predicted probabilities of undecided relative to the other indicators.

In addition to calculating and inspecting predicted probabilities, we can calculate effects of given changes in the latent variable on predicted probabilities of responses for the indicators. Such estimates are sometimes referred to as marginal effects (Long, 1997; Long and Freese, 2014). In this context, a marginal effect is calculated as the difference in the predicted probability of a given response of a given indicator across two different values of a latent variable (or multiple latent variables if an indicator loads on more than one latent variable) —that is, a discrete change of a specific size (Long and Freese, 2014). For our Christian nationalism measurement model, we can write this as

$$me(CN = a, CN = b) = \Pr(x_j = c | CN = a) - \Pr(x_j = c | CN = b), \tag{6.11}$$

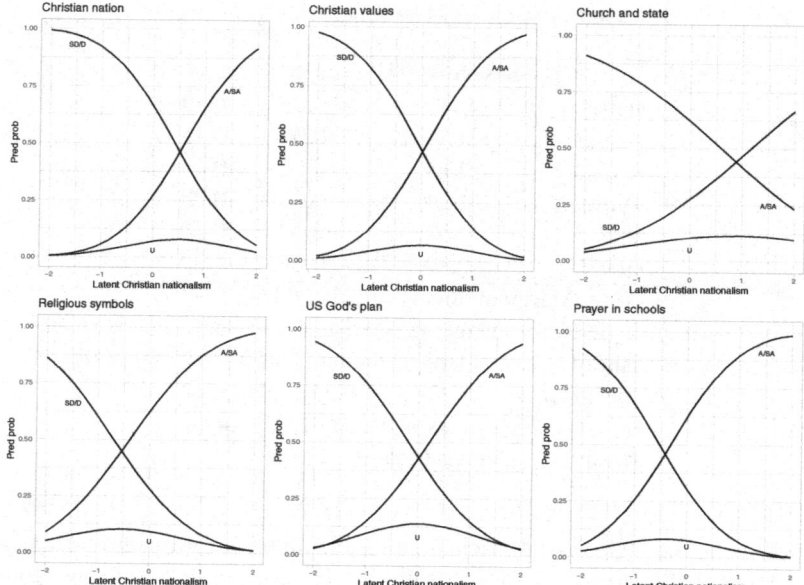

Figure 6.5: Predicted probabilities of responses on indicators of Christian nationalism. *Note:* Pred Prob = predicted probability.

where CN is latent Christian nationalism, a and b are two values for the latent variable, j indexes the indicators, c indexes the responses, and the predicted probabilities are calculated as described above.

Table 6.6 reports estimates of marginal effects based on a change in Christian nationalism from -0.45 to 0.45 (i.e., a 1 standard deviation change centered on the mean of 0). We see that over this range, Christian nationalism has the largest effect on the predicted probability of agreeing for *values* measures and the smallest effect on *separate* (in the opposite direction as per the wording of the question). It is worth noting that this pattern is consistent with the sizes of the unstandardized and standardized loadings, but this will not always be the case. We could calculate marginal effects for other ranges of latent Christian nationalism or, similar to the predicted probabilities, graph how the marginal effects change with a fixed starting point and a range of ending points.

Although examining marginal effects can be quite valuable for interpreting measurement models with categorical indicators, a limitation lies in the relative difficulty in calculating standard errors for the estimates. The delta

Table 6.6: Estimates of selected marginal effects.

	Disagree	Undecided	Agree
nation	−0.32	0.03	0.29
values	−0.35	0.00	0.35
separate	0.15	0.02	−0.18
symbols	−0.25	−0.03	0.28
plan	−0.30	0.00	0.30
prayer	−0.30	−0.02	0.32

Note: Marginal effects calculated based on a discrete change in Christian nationalism of one standard deviation centered on the mean (i.e., −0.45 to 0.45). Marginal effects calculated using the WSL-MV delta parameterization estimates.

method is one approach to calculating standard errors for marginal effects that is available in some statistical software (Xu and Long, 2005), but it can be tedious to program by hand and is not currently an option in statistical software packages designed for structural equation models.

6.4 Comparison With IRT Models

CFA measurement models with ordinal or dichotomous indicators are closely related to the IRT models with differences primarily concerning the conceptualization and parameterization of the models. IRT models emerged as an improvement over classical test theory in evaluating binary items designed to measure latent traits, such as reading ability, and determining how much of a latent trait people have. As such, IRT is oriented more toward identifying levels of latent traits and how properties of items relate to probabilities of responses, while CFA is oriented more toward accounting for the interrelationships among a set of indicators (in particular, categorical indicators in this case). These orientations, however, do not represent sharp distinctions between the two traditions.

In the IRT framework, the latent trait is denoted by θ, and there are broadly one, two, and three parameter models. In the original development, all the items were dichotomous, although a number of IRT models have been extended to account for ordinal items. The one-parameter logistic model

(1PL) or Rasch model (Rasch, 1960) is given by

$$\Pr(x_{ij} = 1) = \frac{\exp(\theta_i - b_j)}{1 + \exp(\theta_i - b_j)}, \tag{6.12}$$

where i indexes individuals, j indexes items, and b_j is the item j's "difficulty." In this parameterization, the item difficulty identifies the level of θ where the probability of response category 1 is 0.5. In the context of a test, this can be interpreted as a measure of how hard the item is. The item difficulty is equivalent to a threshold parameter in a CFA model, and the 1PL model is equivalent to a CFA measurement model in which the factor loadings are fixed to 1.

The two-parameter logistic model (2PL) adds an "item discrimination" parameter. In the context of a test, this parameter is typically interpreted as how well the item differentiates respondents with different levels of ability. The 2PL model is given by

$$\Pr(x_{ij} = 1) = \frac{\exp(a_j(\theta_i - b_j))}{1 + \exp(a_j(\theta_i - b_j))}, \tag{6.13}$$

where a is the discrimination parameter, which is analogous to a factor loading. This model is equivalent to a standard CFA measurement model.

Finally, the three-parameter logistic model (3PL) adds a "guessing" parameter that, again in the context of a test, reflects the possibility that respondents may be able to get answers correct by guessing. The 3PL model can be written as

$$\Pr(x_{ij} = 1) = c_j + (1 - c_j)\frac{\exp(a_j(\theta_i - b_j))}{1 + \exp(a_j(\theta_i - b_j))}, \tag{6.14}$$

where c is the guess parameter. Unlike the previous two IRT models, this model has no direct analogue with a CFA measurement model.

As we can see, the 1PL and 2PL IRT models can be reparameterized to match CFA measurement models. The difference between IRT and CFA thus lies primarily in the interpretation of the parameters, the estimators used (WLS-MV is more common with CFA, while ML with a logit link is more common with IRT), and the research contexts in which the models are used.

6.5 Conclusion

In this chapter, we provided an overview of working with categorical indicators in CFA measurement models. Explicitly accounting for the ordinal (or dichotomous) nature of indicators introduces some complications in assessing model fit and interpreting parameter estimates that we have discussed. The increasingly wide availability of statistical software that include estimators for CFA measurement models with categorical indicators and improvements in computational time, however, make fitting such models reasonably practical, and they should be considered more often.

6.6 Further Reading

Readers interested in learning more about the WLS family of estimators and ML estimators for categorical indicators should see Bovaird and Koziol (2012). For a complete treatment of IRT models, see de Ayala (2009) and Wilson (2005). Long (1997) and Long and Freese (2014) provide excellent discussions of marginal effects for the general class of statistical models with categorical outcomes. For an empirical example exploring measures of material hardship with categorical indicators, see Heflin et al. (2009).

CHAPTER 7. CONCLUSION

CFA provides an elegant, theory-driven approach to measuring concepts in social and behavioral science research. A CFA measurement model specifies the relationships between latent variables representing concepts and indicators of these latent variables with the potential to account for any remaining shared variance in the indicators through correlated errors. Such a model has widespread utility in assessing the psychometric properties of measurement instruments, exploring reliability and validity, and examining the performance of indicators across subpopulations. In addition, developing a measurement model for a concept can be a first step in a larger analysis involving the concept.

7.1 Advanced Topics in CFA

In this book, we have provided researchers a broad foundation in the fundamentals of CFA measurement models. There are a number of more advanced topics that may be of interest to readers in specific contexts. In this section, we provide a brief discussion of four topics and a few pointers for where to start for additional information.

7.1.1 Higher Order and Bifactor Models

The examples we have presented in the book have all involved first-order latent variables in which the latent variables are directly measured by observed indicators. In some cases, however, it is of theoretical or substantive interest to specify a measurement model in which some latent variables are measured by other latent variables. For instance, a model for coping styles might distinguish problem-focused strategies and emotion-focused strategies. These strategies, in turn, could be measured by different dimensions (e.g., problem solving and cognitive restructuring for the problem-focused strategies) that are themselves latent constructs measured by multiple observed indicators. Such a measurement model would be considered a second-order measurement model with the first-order latent variables represented by the specific dimensions and the second-order latent variables represented by the different broader strategies for coping.

In some research contexts, instead of specifying a higher order measurement model, it may make more sense to specify a measurement model in which a focal latent variable has direct relationships with the observed indicators and additional latent variables are included to capture other sources of systematic variation in the observed indicators. In this type of measurement model, referred to as a bifactor model, the focal latent variable is specified as uncorrelated with the additional latent variables. A classic example comes from a study by Holzinger and Swineford (1939) that analyzed 26 measures of tests given to seventh and eighth graders and found one general latent construct to account for ability across all the tests and four domain-specific latent constructs to account for additional systematic variation in the test scores. Brown (2015) provides a discussion of higher order and bifactor models.

7.1.2 Complex Survey Designs

Throughout our text, we have assumed that we have a simple random sample or that departures from that in terms of clustering of cases due to sampling can be addressed with a robust estimator. In many cases, however, analysts work with data from a complex sampling procedure in which respondents had unequal probabilities of being a member of the sample and the sampling design involved stratification and clustering. Data from complex samples typically include weights and variables that capture the sampling design. In general, it is not difficult to incorporate sample weights and information about the sample design into CFA, although there are a couple complications with model fit statistics. Most statistical software packages that fit CFA measurement models have options for including weights and adjusting for the sample design. Muthén and Satorra (1995) provide an overview geared toward structural equation models with CFA as a special case.

7.1.3 Multilevel and Longitudinal CFA

In some cases, the clustered or nested structure of the data is itself of substantive interest rather than a nuisance that needs to be addressed. A canonical example involves having data on, for instance, indicators of school engagement for students nested in classrooms. Multilevel CFA permits the specification of a measurement model for school engagement that partitions the variance of the indicators into a within-classroom component and a between-classroom component. Doing so allows analysts to explore how the measurement of school engagement differs within as compared with between classrooms and to examine the intraclass correlation coefficients for each of the indicators.

Longitudinal data in which multiple observations for the same cases are observed over time are a notable form of multilevel data. In addition to the opportunities available with multilevel data in general, a particular focus with longitudinal CFA lies in exploring measurement invariance over time.

Muthén (1994) provides an overview of working with multilevel data and Meredith and Horn (2001) is a good starting point for understanding measurement invariance in a longitudinal context.

7.1.4 Bayesian Estimators

In this text, we have covered three of the most popular estimators for CFAs: ML and variants for continuous indicators, WLS-MV for categorical indicators, and ML for categorical indicators. In addition, we provided a brief overview of MIIV estimators that have have a number of attractive properties relative to the more popular ML estimators. Another alternative, Bayesian estimators, has received increasing attention for CFA measurement models.

Bayesian estimators begin with the equation

$$\Pr(\theta|x) \propto L(x|\theta)\Pr(\theta), \tag{7.1}$$

where, in this context, θ refers to the set of CFA measurement model parameters and x refers to the data. This simple equation indicates that the *posterior distribution* of the parameters given the data $(\Pr(\theta|x))$ is proportional to the likelihood of the data given the parameters $(L(x|\theta))$ multiplied by the *prior distribution* of the parameters. The likelihood for CFAs can be defined as with ML estimators. The ability to incorporate prior information about the parameters into the analysis has traditionally been a source of skepticism, but the ability to specify *noninformative* or *weakly informative* priors in combination with benefits of Bayesian estimators for CFAs has led to a reconsideration.

The potential benefits of Bayesian estimators relative to popular alternatives include better performance in small samples, a reduced likelihood of returning inadmissible parameter estimates (e.g., negative estimates of error variances), increased statistical power in some settings, and an ability to specify more complex measurement models that can directly account for idiosyncratic features of the data or analysis, among others. Muthén and Asparouhov (2012) and Kaplan and Depaoli (2012) provide overviews and Lee (2007) provides a more technical introduction for readers interested in learning more about Bayesian estimators for CFA.

7.2 Moving Beyond CFA

In some research contexts, CFA is an end in itself, while in others it is only the beginning of the analysis. In developing an analysis beyond CFA, researchers have two broad choices. One option is to integrate measurement models for latent variables into a broader structural equation model that permits the specification of structural relationships among sets of latent and observed variables. Such an approach allows researchers to account for measurement error in their analyses of structural relationships. Bollen (1989) is a classic introduction to the broader SEM framework.

The other option is to use the parameter estimates from a measurement model to predict values for latent variables, which are typically referred to as factor scores. There are several approaches to computing factor scores, although in practice the different approaches result in factor scores that are typically quite similar (Bollen, 1989). Once computed, factor scores can then be used as an observed variable that represents an approximation of an underlying latent variable in further analyses. Although not as principled as structural equation models, factor scores are often highly correlated with an underlying latent variable and thus represent good approximations when used as a proxy in later analyses.

CHAPTER 8. APPENDIX: RELIABILITY OF SCALES

One use of CFA involves obtaining parameter estimates that permit analysts to calculate a more general measure of reliability than the often-used Cronbach's alpha (Cronbach, 1951). Cronbach's alpha assumes that (1) the indicators load on a single latent variable, (2) the factor loadings for the indicators are all equal, (3) there are no correlations among the errors for the indicators, and (4) all the indicators are continuous. In most cases, these assumptions can be relaxed for a set of indicators with a CFA measurement model, and alternative measures of reliability can be calculated. Below, we present three alternatives for continuous indicators and one alternative for categorical indicators. There are a number of other alternatives for various specialized situations, but these four cover a broad range of research contexts and represent an improvement over reporting Cronbach's alpha.

8.0.1 Omega Total

Following the terminology in McNeish (2018), the first alternative measure of scale reliability we consider is *Omega total*. Omega total provides an estimate of reliability for a scale constructed using equal weights (e.g., a summative scale or average scale). Omega total relaxes the assumption of equal factor loadings but maintains the remaining three assumptions of Cronbach's alpha. This measure of reliability based on K indicators of a single latent variable is easily calculated as

$$\omega_{total} = \frac{\left(\sum_{i=1}^{K} \lambda_i\right)^2}{\left(\sum_{i=1}^{K} \lambda_i\right)^2 + \sum_{i=1}^{K} \theta_{ii}}, \tag{8.1}$$

where λ_k is the factor loading (either unstandardized or standardized) and θ_{kk} is the error variance for the kth indicator. Note that this measure of reliability is equivalent to Cronbach's alpha when the factor loadings are equal.

8.0.2 Hancock's Coefficient H

If instead a researcher constructs a scale that weights the observed indicators based on estimated factor loadings rather than using equal weights, then *Hancock's Coefficient H* or simply *H* provides an estimate of reliability (Hancock and Mueller, 2001). As with Omega total, this measure relaxes the assumption of equal factor loadings but maintains the other three assumptions of Cronbach's alpha. H is calculated as

$$H = \left(1 + \left(\sum_{i=1}^{K} \frac{l_i^2}{1 - l_i^2}\right)^{-1}\right)^{-1}, \tag{8.2}$$

where l_k is the standardized factor loading for the kth indicator.

8.0.3 Omega Total Extended

Omega total can be extended to relax the assumption that there are no correlations among the errors for the indicators by including an additional term in the denominator

$$\omega_{tcov} = \frac{\left(\sum_{i=1}^{K} \lambda_i\right)^2}{\left(\sum_{i=1}^{K} \lambda_i\right)^2 + \sum_{i=1}^{K} \theta_{ii} + 2 \sum_{i=2}^{K} \sum_{j=1}^{i} \theta_{ij}}, \tag{8.3}$$

where θ_{ij} captures the covariance between the ith and jth indicators' errors. As with Omega total, this measure of reliability is appropriate for scales constructed using equal weights for the indicators.

8.0.4 Omega for Categorical Indicators

The last measure of reliability we present relaxes the assumption of having continuous indicators and equal factor loadings but maintains the assumptions of loading on a single latent variable and having no correlations among the indicators' errors (Yang and Green, 2015). This measure is an analogue of Omega total for continuous indicators and is appropriate for a scale constructed using equal weights based on dichotomous or ordinal indicators. The measure is calculated as

$$\omega_{cat} = \frac{\sum_{i=1}^{K} \sum_{j=1}^{K} \left[\sum_{d=1}^{C-1} \sum_{e=1}^{C-1} \Phi_2(\tau_{id}, \tau_{je}, l_i l_j) - \left(\sum_{d=1}^{C-1} \Phi_1(\tau_{id})\right)\left(\sum_{e=1}^{C-1} \Phi_1(\tau_{je})\right)\right]}{\sum_{i=1}^{K} \sum_{j=1}^{K} \left[\sum_{d=1}^{C-1} \sum_{e=1}^{C-1} \Phi_2(\tau_{id}, \tau_{je}, \rho_{ij}) - \left(\sum_{d=1}^{C-1} \Phi_1(\tau_{id})\right)\left(\sum_{e=1}^{C-1} \Phi_1(\tau_{je})\right)\right]}, \tag{8.4}$$

where k indexes indicators and c indexes categories, τ_{kc} references threshold c for indicator k, l_k is the standardized factor loadings from a WLS-MV estimator for indicator k, and ρ_{ij} is the polychoric correlation between indicators

i and j. The two functions Φ_1 and Φ_2 refer to the univariate and bivariate standard normal distribution functions. This equation is a bit more difficult to use for hand calculations, but it can be written as a function in various statistical software packages.

CHAPTER 9. GLOSSARY

Akaike Information Criterion (AIC): An index used to assess the fit of a given measurement model relative to an alternative, potentially nonnested, measurement model. The AIC applies a penalty based on the number of parameters estimated.

Bayesian Information Criterion (BIC): An index used to assess the fit of a given measurement model relative to an alternative, potentially nonnested, measurement model. The BIC applies a stronger penalty than the AIC based on the number of parameters estimated.

Causal (Formative) Indicator: An indicator that is the *cause of* a latent variable.

Communality: The amount of variance in an indicator explained by the latent variables.

Comparative Fit Index (CFI): An index for assessing the overall fit of a measurement model. The CFI ranges from 0 to 1 with values above 0.95 usually taken to be indicative of good fit. Some versions of the CFI permit values greater than 1.

Configural Invariance: A form of measurement invariance in which the model specification is equivalent across a set of groups or subpopulations.

Delta Parameterization: An identification strategy for models with categorical indicators that involves fixing the variances of the underlying continuous variables representing the observed categorical variables to 1.

Differential Item Functioning (DIF): A special case of lack of measurement invariance where an indicator (item) performs differentially for two or more groups at the same level of the latent variable; either by differences in means (uniform DIF) or differences in means and factor loadings (nonuniform DIF).

Effect (Reflective) Indicator: An indicator that is *caused by* a latent variable.

Endogenous Variable: A "dependent" variable or a variable that is determined by other variables in the measurement model.

Exogenous Variable: An "independent" variable or a variable that is not determined by other variables in the model.

Factor Loading: A parameter of a measurement model that reflects the relationship between a latent variable and an indicator of the latent variable.

Implied Moment Matrices: The specification of a measurement model implies a structure among the means, variances, and covariances of a set of observed indicators. Implied moment matrices refer to this structure and can be compared with the actual means, variances, and covariances among a set of observed indicators to assess the fit of the model.

Indicator: A observed (manifest) variable that serves as measure of a latent variable.

Intercept: A parameter of a measurement model that reflects the mean value of an indicator when a latent variable takes the value of 0.

Item Response Theory (IRT): IRT is a methodological approach that emerged as an improvement over classical test theory in evaluating binary items designed to measure latent traits and determining how much of a latent trait people have.

Latent Variable: An unobserved variable that represents a concept and is measured by observed indicators.

Logit: The logarithm of the odds for a probability. It is one of the link functions commonly used in estimators for categorical indicators.

Mean- and Variance-Adjusted Weighted Least Squares (WLS-MV): An estimator for categorical indicators that addresses the computational challenges of the WLS estimator.

Measurement Invariance: A quality of a measurement model that refers to the equivalence of the model structure and parameters across groups or subpopulations.

Measurement Model: A model that relates one or more latent variables to observed indicators (measures) of each of the latent variables.

Metric Invariance: A form of measurement invariance in which the factor loadings for all the indicators are equal across a set of groups or subpopulations.

Multiple-Indicator Multiple-Cause (MIMIC) Model: A model used to assess uniform and nonuniform differential item functioning.

Maximum Likelihood: A broad class of estimators based on maximizing a likelihood function defined by a probability distribution for a set of observed indicators.

Polychoric Correlation Matrix: A matrix of correlation coefficients based on underlying continuous variables for a set of observed binary or ordinal variables.

Probit Function: The quantile function associated with a standard normal distribution. It is one of the link functions commonly used in estimators for categorical indicators.

Root Mean Square Error of Approximation (RMSEA): An index for assessing the overall fit of a measurement model. The RMSEA has a minimum value of 0 and no upper bound. Values below 0.05 are usually taken to be indicative of good model fit and values below 0.10 as adequate model fit.

Scalar Invariance: A form of measurement invariance in which the intercepts for all the indicators are equal across a set of groups or subpopulations.

Schwarz-Modified Bayesian Information Criterion (SBIC): A variant of the BIC that has a convenient form that permits an assessment of overall model fit in addition to relative model fit. Models with SBIC < 0 fit the data better than a saturated model that permits all indicators to be correlated with each other.

Standardized Root Mean Residuals (SRMR): An index for assessing the overall fit of a measurement model based on the differences between the implied covariance matrix and the observed covariance matrix. The SRMR has a range of 0 to 1 with values below 0.08 usually taken to be indicative of good fit.

Theta Parameterization: An identification strategy for models with categorical indicators that involves fixing the variances of the indicator errors to 1.

Threshold (cut-point): A parameter of a measurement model with categorical indicators. Thresholds (or cut-points) represent the points on a latent continuum at which the probability of a response of a given value shifts to the next value (e.g., moving from 2 to 3 on a Likert-type scale).

Tucker–Lewis Index (TLI): An index for assessing the overall fit of a measurement model. The TLI ranges from 0 to 1 with values above 0.95 usually taken to be indicative of good fit. Some versions of the TLI permit values greater than 1.

Weighted Least Squares (WLS): A broad class of estimators that was used in the CFA context for nonnormal or categorical indicators. The poor performance of the standard WLS estimator led to the development of variants, such as the WLSMV and DWLS estimators.

REFERENCES

Agresti, A. (2002). *Categorical Data Analysis, 2nd ed.* John Wiley, Hoboken, NJ.

Akaike, H. (1973). Information theory and an extension of the maximum likelihood principle. In Petrov, B. N. and Csaki, F., editors, *Second International Symposium on Information Theory*, pages 267–281. Akademiai Kiado, Budapest.

Alwin, D. F. (2007). *Margins of Error: A Study of Reliability in Survey Measurement.* John Wiley, Hoboken, NJ.

Arbuckle, J. L. (1996). Full information estimation in the presence of missing data. In Marcoulides, G. A. and Schumacker, R. E., editors, *Advanced Structural Equation Modeling: Issues and Techniques*, pages 243–277. Erlbaum, Mahwah, NJ.

Bauldry, S. (2014). miivfind: A command for identifying model-implied instrumental variables for structural equation models in stata. *Stata Journal*, 14:60–75.

Bentler, P. M. (1990). Comparative fit indexes in structural equation models. *Psychological Bulletin*, 107:238–246.

Bentler, P. M. (1995). *EQS Structural Equations Program Manual.* Multivariate Software, Encino, CA.

Bollen, K. A. (1982). A confirmatory factor analysis of subjective air quality. *Evaluation Review*, 6:521–535.

Bollen, K. A. (1989). *Structural Equations With Latent Variables.* John Wiley, New York, NY.

Bollen, K. A. (1996). An alternative two stage least squares (2SLS) estimator for latent variable equations. *Psychometrika*, 61:109–121.

Bollen, K. A. (2001). Two-stage least squares and latent variable models: simultaneous estimation and robustness to misspecification. In Cudeck, R., Toit, S. T., and S orbom, D., editors, *Structural Equation Modeling: Present and Future, a Festschrift in Honor of Karl J oreskog*, pages 119–138. Scientific Software International, Lincolnwood, IL.

Bollen, K. A. and Bauldry, S. (2010). Model identification and computer algebra. *Sociological Methods & Research*, 39:127–156.

Bollen, K. A. and Bauldry, S. (2011). Three cs in measurement models: Causal indicators, composite indicators, and covariates. *Psychological Methods*, 16:264–284.

Bollen, K. A. and Lennox, R. D. (1991). Conventional wisdom on measurement: A structural equation perspective. *Psychological Bulletin*, 110:305–314.

Bovaird, J. A. and Koziol, N. A. (2012). Measurement models for ordered categorical indicators. In Hoyle, R. H., editor, *Handbook of Structural Equation Modeling*, pages 495–511. Guilford Press, New York, NY.

Brown, T. A. (2015). *Confirmatory Factor Analysis for Applied Research. 2nd ed.* Multivariate Software, Encino, CA.

Browne, M. W. and Cudeck, R. (1993). Alternative ways of assessing model fit. In Bollen, K. A. and Long, J. S., editors, *Testing Structural Equation Models*, pages 136–162. Sage, Newbury Park, CA.

Byrne, B. M., Shavelson, R. J., and Muthén, B. (1989). Testing for the equivalence of factor covariance and mean structures: The issue of partial measurement invariance. *Psychological Bulletin*, 105:456–466.

Campbell, D. T. and Fiske, D. W. (1959). Convergent and discriminant validation by the multitrait-multimethod matrix. *Psychological Bulletin*, 56:81–105.

Cieciuch, J., Davidov, E., Algesheimer, R., and Schmidt, P. (2018). Testing for approximate measurement invariance of human values in the european social survey. *Sociological Methods & Research*, 47:665–686.

114

Cronbach, L. J. (1951). Coefficient alpha and the internal structure of tests. *Psychometrika*, 16:297–334.

Curran, P. J., Bollen, K. A., Chen, F., Paxton, P., and Kirby, J. B. (2003). Finite sampling properties of the point estimates and confidence intervals of the rmsea. *Sociological Methods & Research*, 32:208–252.

de Ayala, R. J. (2009). *The Theory and Practice of Item Response Theory*. Guildford Press, New York, NY.

Eliason, S. R. (1993). *Maximum Likelihood Estimation: Logic and Practice*. Sage, Newbury Park, CA.

Enders, C. K. (2010). *Applied Missing Data Analysis*. Guilford Press, New York, NY.

Finch, W. H. (2020). *Exploratory Factor Analysis*. Sage, Thousand Oaks, CA.

Fisher, Z., Bollen, K. A., Gates, K., and Rönkkö, M. (2019). *Model Implied Instrumental Variable (MIIV) Estimation of Structural Equation Models*. R package Version 0.5.4.

Flora, D. B. and Curran, P. J. (2004). An empirical evaluation of alternative methods of estimation for confirmatory factor analysis with ordinal data. *Psychological Methods*, 9:466–491.

Fox, J. (2016). *Applied Regression Analysis & Generalized Linear Models, 3rd ed.* Sage, Thousand Oaks, CA.

Fullerton, A. (2009). A conceptual framework for ordered logistic regression models. *Sociological Methods & Research*, 38:306–347.

Goertz, G. (2020). *Social Science Concepts and Measurement*. Princeton University Press, Princeton, NJ.

Gujarati, D. N. (2018). *Linear Regression: A Mathematical Introduction*. Sage, Thousand Oaks, CA.

Hancock, G. R. and Mueller, R. O. (2001). Rethinking construct reliability within latent variable systems. In Cudeck, R., du Toit, S., and Sörbom, D., editors, *Structural Equation Modeling: Present and Future*, pages 195–216. Scientific Software International, Lincolnwood, IL.

Harris, K. M., Halpern, C. T., Whitsel, E., Hussey, J., Tabor, J., Entzel, P., and Udry, J. R. (2009). The National Longitudinal Study of Adolescent Health: Research Design. Technical report.

Heflin, C., Sandberg, J., and Rafail, P. (2009). The structure of material hardship in u.s. households: An examination of the coherence behind common measures of well-being. *Social Problems*, 56:746–764.

Hempel, L. M., Matthews, T., and Bartkowski, J. (2012). Trust in a "fallen world": The case of protestant theological conservatism. *Journal for the Scientific Study of Religion*, 51:522–541.

Holzinger, K. J. and Swineford, F. (1939). A study in factor analysis: The stability of a bi-factor solution. *Supplementary Educational Monographs*, 48:1–91.

Hu, L.-t. and Bentler, P. M. (1999). Cutoff criteria for fit indexes in covariance structure analysis: Conventional criteria versus new alternatives. *Structural Equation Modeling*, 6:1–55.

Iacobucci, D. (2008). *Mediation Analysis*. Sage, Thousand Oaks, CA.

Inglehart, R. C., Haerpfer, A. M., Welzel, C., Kizilova, K., Diez-Medrano, J., Lagos, M., Norris, P., Ponarin, E., and Puranen, B. (2014). *World Values Survey: Round Six—Country-Pooled Datafile 2010-2014*. JD Systems Institute, Madrid.

Jöreskog, K. and Goldburger, A. S. (1975). Estimation of a model with multiple indicators and multiple causes of a single latent variable. *Journal of the American Statistical Association*, 70:631–639.

Kaplan, D. and Depaoli, S. (2012). Bayesian structural equation modeling. In Hoyle, R., editor, *Handbook of Structural Equation Modeling*, pages 650–673. Guilford Press, New York, NY.

Kessler, R. C., Andrews, G., Colpe, L. J., Hiripi, E., Mroczek, D. K., Normand, S.-L. T., Walters, E. E., and Zaslavsky, A. M. (2002). Short screening scales to monitor population prevalences and trends in non-specific psychological distress. *Psychological Medicine*, 32:959–976.

Kline, R. B. (2015). *Principles and Practice of Structural Equation Modeling, 4th ed.* Guilford Press, New York, NY.

Kolenikov, S. and Bollen, K. A. (2012). Testing negative error variances: Is a Heywood case a symptom of misspecification. *Sociological Methods & Research*, 41:124–167.

Lee, S. (2007). *Structural Equation Modeling: A Bayesian Approach.* John Wiley, New York, NY.

Lewis-Beck, C. and Lewis-Beck, M. (2015). *Applied Regression: An Introduction, 2nd ed.* Sage, Thousand Oaks, CA.

Little, R. J. A. and Rubin, D. B. (2019). *Statistical Analysis With Missing Data, 3rd ed.* John Wiley, Hoboken, NJ.

Long, J. S. (1997). *Regression Models for Categorical and Limited Dependent Variables.* Sage, Thousand Oaks, CA.

Long, J. S. and Freese, J. (2014). *Regression Models for Categorical Dependent Variables Using Stata, 3rd ed.* Stata Press, College Station, TX.

Lord, F. M., Novick, M. R., and Birnbaum, A. (1968). *Statistical Theories of Mental Test Scores.* Addison-Wesley, Oxford.

MacCallum, R. C., Roznowski, M., and Necowitz, L. B. (1992). Model modifications in covariance structure analysis: The problem of capitalization on chance. *Psychological Bulletin*, 111:490–504.

Manglos-Weber, N. D., Mooney, M. A., Bollen, K. A., and Roos, J. M. (2016). Relationships with god among young adults: Validating a measurement model with four dimensions. *Sociology of Religion*, 77:193–213.

McCutcheon, A. L. (1987). *Latent Class Analysis.* Sage, Thousand Oaks, CA.

McNeish, D. (2018). Thanks coefficient alpha, we'll take it from here. *Psychological Methods*, 23:412–433.

Meredith, W. (1993). Measurement invariance, factor analysis and factorial invariance. *Psychometrika*, 58:525–543.

Meredith, W. and Horn, J. (2001). The role of factorial invariance in modeling growth and change. In Collins, L. M. and Sayer, A. G., editors, *New Methods for the Analysis of Change*, pages 203–240. American Psychological Association, Washington, D.C.

Mize, T. (2019). Doing gender by criticizing leaders: Public and private displays of status. *Social Problems*, 66:86–107.

Muthén, B. O. (1985). A method for studying the homogeneity of test items with respect to other relevant variables. *Journal of Educational Statistics*, 10:121–132.

Muthén, B. O. (1994). Multilevel covariance structure analysis. *Sociological Methods & Research*, 22:376–398.

Muthén, B. O. and Asparouhov, T. (2012). Bayesian structural equation modeling: A more flexible representation of substantive theory. *Psychological Methods*, 17:313–335.

Muthén, B. O. and Satorra, A. (1995). Complex sample data in structural equation modeling. *Sociological Methodology*, 25:267–316.

Muthén, L. K. and Muthén, B. O. (2017). *Mplus User's Guide, 8th edition.* Muthén & Muthén, Los Angeles, CA.

Olsson, U. (1979). Maximum likelihood estimation of the polychoric correlation coefficient. *Psychometrika*, 44:443–460.

Osterlind, S. J. and Everson, H. T. (2009). *Differential Item Functioning.* Sage, Thousand Oaks, CA.

Ostini, R. and Nering, M. L. (2006). *Polytomous Item Response Theory Models*. Sage, Thousand Oaks, CA.

Paxton, P., Hipp, J. R., and Marquart-Pyatt, S. (2011). *Nonrecursive Models: Endogeneity, Reciprocal Relationships, and Feedback Loops*. Sage, Thousand Oaks, CA.

Preacher, K. J., Wichman, A. L., MacCallum, R. C., and Briggs, N. E. (2008). *Latent Growth Curve Modeling*. Sage, Thousand Oaks, CA.

Raftery, A. E. (1995). Bayesian model selection in social research. *Sociological Methodology*, 25:111–163.

Rasch, G. (1960). *Probabilistic Models for Some Intelligence and Attainment Tests*. University of Chicago Press, Chicago.

Rhemtulla, M., Brosseau-Liard, P. É., and Savalei, V. (2012). When can categorical variables be treated as continuous? A comparison of robust continuous and categorical sem estimation methods under suboptimal conditions. *Psychological Methods*, 17:354–373.

Roos, J. M. (2014). Measuring science or religion? a measurement analysis of the national science foundation sponsored science literacy scale 2006–2010. *Public Understanding of Science*, 23:797–813.

Roos, J. M., Hughes, M., and Reichelmann, A. V. (2019). A puzzle of racial attitudes: A measurement analysis of racial attitudes and policy indicators. *Socius*, 5:Advance online publication.

Rothenberg, T. J. (1971). Identification of parametric models. *Econometrica*, 39:577–591.

Satorra, A. (2000). Scaled and adjusted restricted tests in multi-sample analysis of moment structures. In Heijmans, R. D. H., Pollock, D. S. G., and Satorra, A., editors, *Innovations in Multivariate Statistical Analysis. A Festschrift for Heinz Neudecker*, pages 233–247. Kluwer Academic, London.

Satorra, A. and Bentler, P. M. (1994). Corrections to test statistics and standard errors in covariance structure analysis. In von Eye, A. and Clogg, C. C., editors, *Latent Variable Analysis*, pages 399–419. Sage, Thousand Oaks.

Satorra, A. and Bentler, P. M. (2010). Ensuring positiveness of the scaled difference chi-square test statistic. *Psychometrika*, 75:243–248.

Savalei, V. (2014). Understanding robust corrections in structural equation modeling. *Structural Equation Modeling*, 21:149–160.

Silva, B. C., Bosancianu, C. M., and Littvay, L. (2019). *Multilevel structural equation modeling*. Sage, Thousand Oaks, CA.

Steiger, J. H. and Lind, J. C. (1980). Statistically-based tests for the number of common factors. In *Annual Meeting of the Psychometric Society*.

Tucker, L. R. and Lewis, C. (1973). A reliability coefficient for maximum likelihood factor analysis. *Psychometrica*, 38:1–10.

van Schuur, W. H. (2011). *Ordinal Item Response Theory: Mokken Scale Analysis*. Sage, Thousand Oaks, CA.

West, S. G., Taylor, A. B., and Wu, W. (2012). Model fit and model selection in structural equation modeling. In Hoyle, R. H., editor, *Handbook of Structural Equation Modeling*, pages 209–231. Guilford Press, New York, NY.

Whitehead, A. L., Perry, S. L., and Baker, J. O. (2018). Make america christian again: Christian nationalism and voting for donald trump in the 2016 election. *Sociology of Religion*, 79:147–171.

Wilson, M. (2005). *Constructing Measures: An Item Response Modeling Approach*. Psychology Press, New York, NY.

Woods, C. M. (2009). Evaluation of mimic-model methods for dif testing with comparison to two-group analysis. *Multivariate Behavioral Research*, 44:1–27.

Woods, C. M. and Grimm, K. J. (2011). Testing for nonuniform differential item functioning with multiple indicator multiple cause models. *Applied Psychological Measurement*, 35:339–361.

Xu, J. and Long, J. S. (2005). Confidence intervals for predicted outcomes in regression models for categorical outcomes. *Stata Journal*, 5:537–559.

Yang, Y. and Green, S. (2015). Evaluation of structural equation modeling estimates of reliability for scales with ordered categorical items. *Methodology*, 11:23–34.

Yuan, K.-H. and Bentler, P. M. (2000). Three likelihood-based methods for mean and covariance structure analysis with nonnormal missing data. *Sociological Methodology*, 30:165–200.

INDEX

120

122